United States Nuclear Regulatory Commission

Protecting People and the Environment

FISCAL YEAR 2010
PERFORMANCE AND ACCOUNTABILITY REPORT

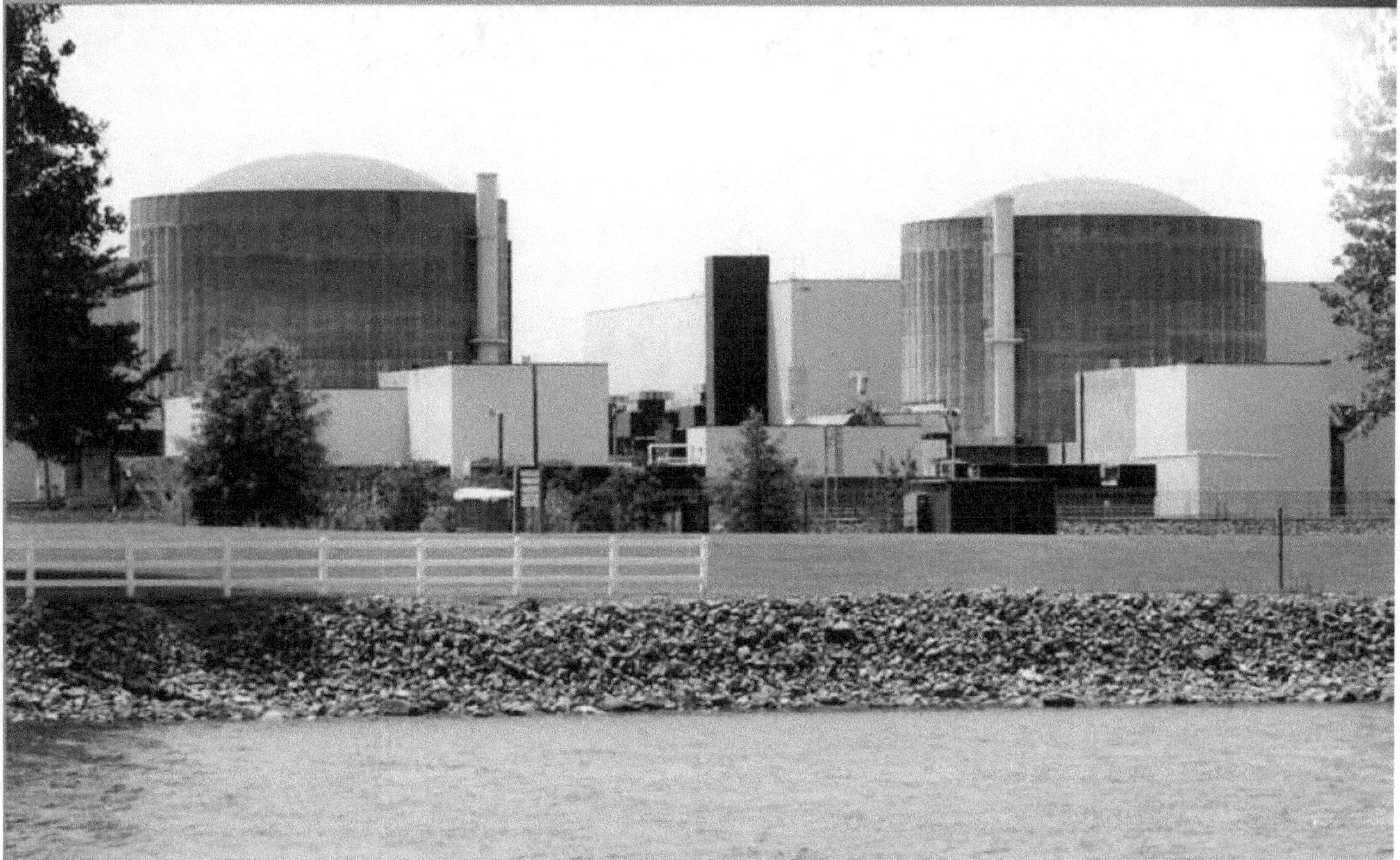

McGuire Nuclear Power Plant, Mecklenburg County, NC

MISSION

License and regulate the Nation's civilian use of

byproduct, source, and special nuclear materials

to ensure adequate protection of public health

and safety, promote the common defense and

security, and protect the environment.

Paperwork Reduction Act Statement
The information collections contained in this document are subject to the Paperwork Reduction Act of 1995 (44 U.S.C. 3501 et seq.), which were approved by the Office of Management and Budget (OMB), approval numbers 3150-0002, 3150-0003, 3150-0004, 3150-0009, 3150-0011, 3150-0012, 3150-0014, 3150-0058, 3150-0104, 3150-0123, 3150-0139, and 3150-0197.

Public Protection Notification
The NRC may not conduct or sponsor, and a person is not required to respond to, a request for information or an information collection requirement unless the requesting document displays a currently valid OMB control number.

Table of Contents

This Performance and Accountability Report is available on the NRC Web page at http://www.nrc.gov

Left to right: Commissioner William D. Magwood, IV, Commissioner Kristine L. Svinicki, Chairman Gregory B. Jaczko, Commissioner George Apostolakis, and Commissioner William C. Ostendorff.

The FY 2010 Performance and Accountability Report provides performance results and audited financial statements that enable Congress, the President, and the public to assess the performance of the agency in achieving its mission and stewardship of its resources. The report contains a concise overview, Management's Discussion and Analysis, as well as performance and financial sections. Details of performance results and program evaluations can be found in the Other Accompanying Information section.

A Message from the Chairman

I am pleased to present the U.S. Nuclear Regulatory Commission's (NRC) Performance and Accountability Report for Fiscal Year (FY) 2010. The report provides key financial and performance information to Congress and the American people. The NRC received the Certificate of Excellence in Accountability Reporting from the Association of Government Accountants for the ninth year in a row for our FY 2009 Performance and Accountability Report. The receipt of this prestigious award demonstrates our commitment to accountability and the high quality reporting of performance and financial information.

We also received an unqualified opinion on the agency's financial statements for the seventh consecutive year. The unqualified opinion attests to NRC's sound financial performance over the past year in support of our mission of protecting public health and safety, promoting common defense and security, and protecting the environment in the civilian use of nuclear materials. This report highlights our achievements and challenges in meeting our mission through the agency's two strategic goals of safety and security, while adhering to the principles of good regulation—independence, openness, efficiency, clarity, and reliability.

In FY 2010, while the NRC maintained effective and efficient oversight of 104 nuclear power plants through emphasis on strengthening the interrelationship among safety, security, and emergency preparedness, the agency also reviewed the critical safety aspects of new reactor designs, environmental siting and combined license applications for the construction of new nuclear power plants. The NRC remained focused on the safe and secure use of nuclear materials through effective oversight of fuel facilities, uranium recovery sites, decommissioning sites, and nuclear material user licensees. In addition, the agency completed significant fuel cycle and materials users licensing reviews and continued reviews of applications for uranium enrichment facilities and uranium recovery to assure protection of public health and safety and the environment.

Commensurate with NRC's achievements and challenges, the NRC is committed to prudently managing the resources entrusted to it by the American people. The NRC continues to evaluate, test, and strengthen its internal controls, including those related to financial reporting and financial management systems, as required by the Federal Managers' Financial Integrity Act (FMFIA). Based on the FMFIA assessments, I have concluded that there is reasonable assurance that the NRC is in substantial compliance with FMFIA, and the financial and performance data published in this report is accurate, reliable and timely. Additionally, I have determined that the agency is in substantial compliance with the Federal Financial Management Improvement Act (FFMIA), based on NRC's application of the FFMIA risk model.

Assuring the public of the agency's commitment to safety and security through openness and transparency is an ongoing challenge. The NRC's Open Government Plan, developed and published in FY 2010, demonstrates the agency's commitment to increasing transparency with the public. The coming year also brings an unprecedented challenge as the agency's operating reactors programs will be subject to peer review by the International Atomic Energy Agency.

The NRC is proud of this year's performance of its 3,981 employees in achieving the agency's safety and security goals and looks forward to continuing its high-quality service to the American people in FY 2011 and beyond.

Gregory B. Jaczko

Gregory B. Jaczko
Chairman
November 12, 2010

2009 Certificate of Excellence

CERTIFICATE OF EXCELLENCE IN ACCOUNTABILITY REPORTING®

Presented to the

U.S. Nuclear Regulatory Commission

In recognition of your outstanding efforts preparing NRC's Performance and Accountability Report for the fiscal year ended September 30, 2009.

A *Certificate of Excellence in Accountability Reporting* is presented by AGA to federal government agencies whose annual Performance and Accountability Reports achieve the highest standards demonstrating accountability and communicating results.

John H. Hummel, CGFM
Chair, Certificate of Excellence
in Accountability Reporting Board

Relmond P. Van Daniker, DBA, CPA
Executive Director, AGA

Chapter 1

Management's Discussion and Analysis

Photo Courtesy of NRC Photo Library

Chairman Gregory B. Jaczko and Executive Director for Operations R. William Borchardt accepting the "Best Place to Work in the Federal Government" honor given by the Partnership for Public Service and the Institute for the Study of Public Policy Implementation.

U.S.NRC
UNITED STATES NUCLEAR REGULATORY COMMISSION

Photo Courtesy of NRC Photo Library

The U.S. Nuclear Regulatory Commission (NRC) headquarters

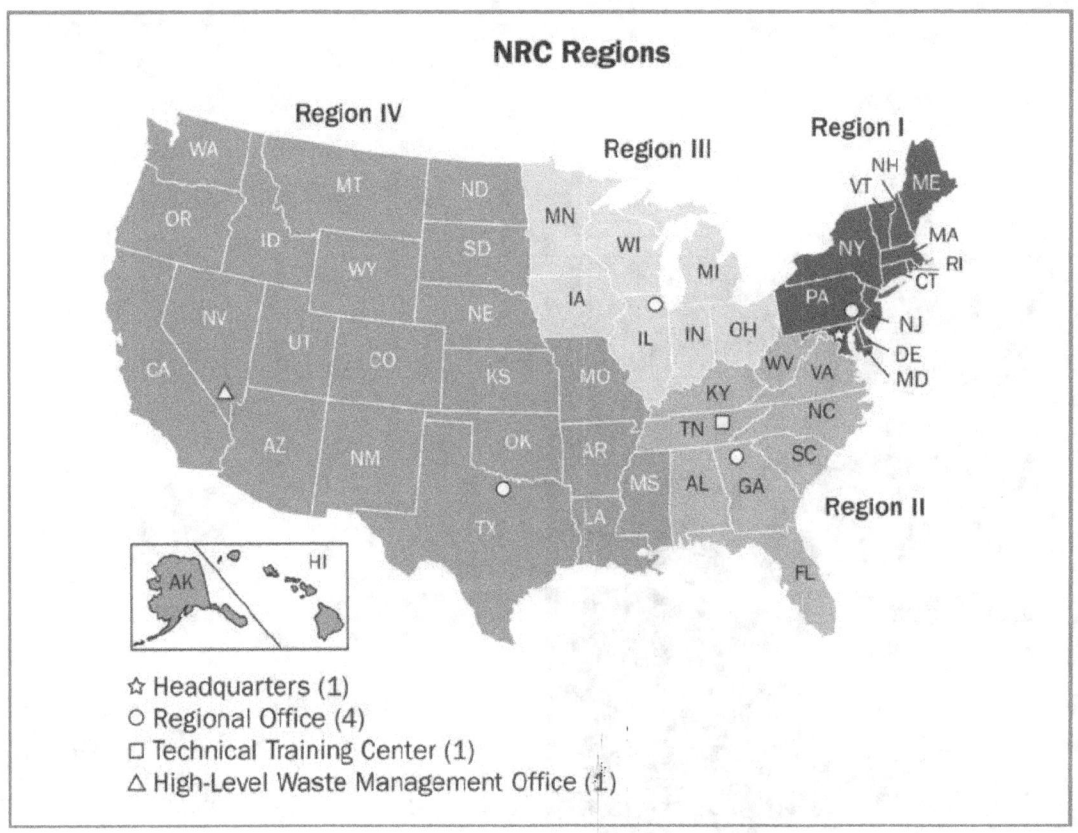

NRC Regions

☆ Headquarters (1)
○ Regional Office (4)
□ Technical Training Center (1)
△ High-Level Waste Management Office (1)

Introduction

The U.S. Nuclear Regulatory Commission (NRC) Performance and Accountability Report presents the agency's program performance and financial management information for fiscal year (FY) 2010. The annual report provides the public with an opportunity to assess how effectively the NRC uses its funds to achieve results. When preparing this report, the NRC staff followed the requirements of the Chief Financial Officers Act, as amended by the Reports Consolidation Act of 1990, Government Management Reform Act of 1994, and Government Performance Results Act of 1993. This Performance and Accountability Report covers activities from October 1, 2009, to September 30, 2010.

The NRC emphasizes keeping the public informed of its activities. Visit the agency's Web site at http://www.nrc.gov to access this report online and to learn more about the NRC and what we do to serve the American public.

Chapter 1, "Management's Discussion and Analysis," provides an overview of the NRC and its accomplishments during FY 2010. Chapter 1 consists of the following six sections: "About the NRC" describes the agency's mission, organizational structure, and regulatory responsibility; "Program Performance Overview" summarizes the agency's success in achieving its strategic goals, which are further described in Chapter 2; "Program Performance Results" outlines the results of the agency's program performance; "Future Challenges" includes forward-looking information; "Financial Performance Overview" highlights the NRC's financial position and audit results contained in Chapter 3; and "Systems, Controls, and Legal Compliance" describes the agency's compliance with key legal and regulatory requirements.

About the NRC

The NRC began operations on January 19, 1975, as an independent Federal agency to regulate the commercial and institutional uses of nuclear materials. The Atomic Energy Act of 1954, as amended, and the Energy Reorganization Act of 1974, as amended, define the NRC's purpose. These acts provide the foundation for the NRC's mission to regulate the Nation's civilian use of byproduct, source, and special nuclear materials to ensure adequate protection of public health and safety, to promote the common defense and security, and to protect the environment.

The agency regulates civilian nuclear power plants, other nuclear facilities, and other uses of nuclear materials. These other uses include nuclear medicine programs at hospitals; academic activities at educational institutions; research work; industrial applications, such as gauges and testing equipment; and the transport, storage, and disposal of nuclear materials and wastes.

To fulfill its responsibility to protect public health and safety, the NRC performs the following three principal regulatory functions:

(1) establishes standards and regulations;

(2) issues licenses for nuclear facilities and users of nuclear materials;

(3) inspects facilities and users of nuclear materials to ensure compliance with regulatory requirements.

Organization

The NRC is headed by a Commission composed of five members, with one member designated by the President to serve as Chairman (see NRC Organizational Chart on page 4). With the advice and consent of the Senate, the President appoints each member to serve a 5-year term. The Chairman is the principal executive officer and official spokesman for the Commission. The Executive Director for Operations carries out policies and decisions made by the Commission, and directs the activities of the programs.

The NRC's Headquarters is located in Rockville, MD. Four regional offices are located in King of Prussia, PA; Atlanta, GA; Lisle, IL; and Arlington, TX. In addition, the NRC's technical training center is located in Chattanooga, TN. The NRC also employs at least two resident inspectors at each of the Nation's 104 nuclear power reactor sites. The NRC's Operations

![U.S.NRC logo] U.S.NRC
UNITED STATES NUCLEAR REGULATORY COMMISSION

U.S. Nuclear Regulatory Commission

![U.S.NRC logo] U.S.NRC
United States Nuclear Regulatory Commission
Protecting People and the Environment

The Commission

Commissioner
William D. Magwood, IV

Commissioner
Kristine L. Svinicki

Chairman
Gregory B. Jaczko

Commissioner
George Apostolakis

Commissioner
William C. Ostendorff

Executive Director, Advisory Committee on Reactor Safeguards
Edwin M. Hackett

Chief Administrative Judge (Chairman), Atomic Safety and Licensing Board Panel
E. Roy Hawkens

Director, Office of Commission Appellate Adjudication
Brooke D. Poole

Director, Office of Congressional Affairs
Rebecca L. Schmidt

Director, Office of Public Affairs
Eliot B. Brenner

Chief Financial Officer
Jim Dyer

Inspector General *
Hubert T. Bell

Secretary of the Commission
Annette L. Vietti-Cook

Director, Office of International Programs
Margaret M. Doane

General Counsel
Stephen G. Burns

Executive Director for Operations
R. William Borchardt

Assistant for Operations
Nader L. Mamish

Deputy Executive Director for Reactor and Preparedness Programs
Martin J. Virgilio

Deputy Executive Director for Materials, Waste, Research, State, Tribal and Compliance Programs
Michael F. Weber

Deputy Executive Director for Corporate Management
Darren B. Ash

Regional Administrator Region I
Bill Dean

Regional Administrator Region II
Luis A. Reyes

Regional Administrator Region III
Mark A. Satorius

Regional Administrator Region IV
Elmo E. Collins

Director, Office of New Reactors
Michael R. Johnson

Director, Office of Nuclear Security and Incident Response
James T. Wiggins

Director, Office of Nuclear Reactor Regulation
Eric J. Leeds

Director, Office of Nuclear Regulatory Research
Brian W. Sheron

Director, Office of Enforcement
Roy P. Zimmerman

Director, Office of Nuclear Material Safety and Safeguards
Catherine Haney

Director, Office of Investigations
Cheryl L. McCrary

Director, Office of Federal and State Materials and Environmental Management Programs
Charles L. Miller

Director, Office of Human Resources
James F. McDermott

Director, Office of Small Business and Civil Rights
Corenthis B. Kelley

Director, Office of Information Services
Thomas M. Boyce

Director, Office of Administration
Kathryn O. Greene

Director, Computer Security Office
Patrick D. Howard

The dotted line signifies that the IG exercises a much higher degree of independence with the Chairman in carrying out his roles and responsibilities in comparison to other executives reporting to the Chairman.

Figure 1

NRC BUDGETARY AUTHORITY, FY 2005–2010

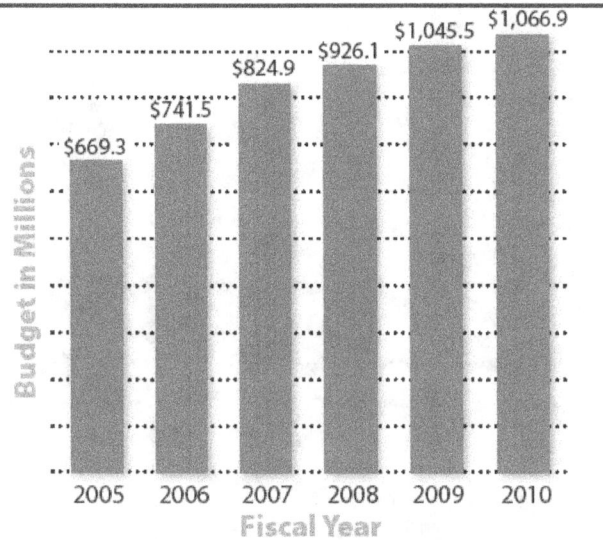

Source: NRC Performance Budget Fiscal Year 2011

Center, located at NRC Headquarters, is the focal point for the agency's communications with its licensees, State agencies, and other Federal agencies about operating events in the commercial nuclear sector. NRC operations officers staff the Operations Center 24 hours a day, 7 days a week.

The NRC's budget for FY 2010 was $1,066.9 million (see Figure 1) with 3,981 full-time equivalent staff (see Figure 2). The NRC recovers approximately 90% of its appropriations from fees paid by NRC licensees and applicants for a license.

The Nuclear Industry

The NRC regulates the commercial use of radioactive materials. The nuclear material cycle begins with the mining and production of nuclear fuel, continues with the use of nuclear fuel to power the Nation's 104 nuclear power plants (see Figure 3, page 6), and ends with the safe transportation and storage of spent nuclear fuel and other nuclear waste. The NRC's regulatory programs ensure that radioactive materials are used safely and securely at every stage in the nuclear material cycle. The NRC oversees 3,000 licenses for medical, academic, industrial, and general uses of nuclear materials.

Figure 2

NRC PERSONNEL CEILING, FY 2005–2010

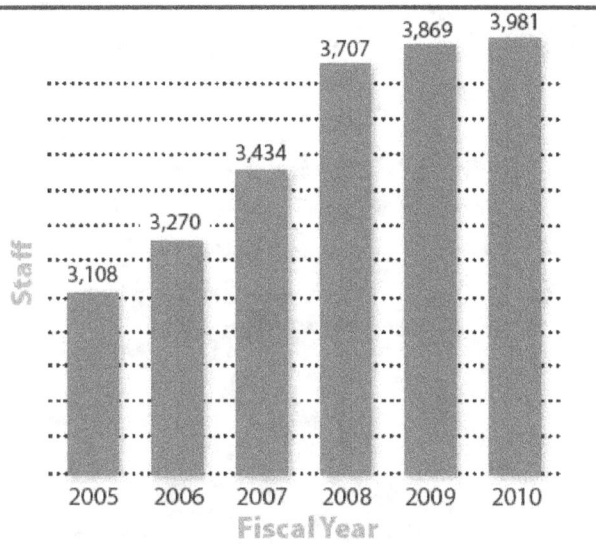

Source: NRC Performance Budget Fiscal Year 2011

The agency conducts approximately 1,200 health and safety inspections of its nuclear materials licensees annually.

Under the NRC's Agreement State program, 37 States have assumed primary regulatory responsibility over the industrial, medical, and other nuclear materials in their States. The NRC works closely with these States to ensure that they maintain public safety consistent with NRC standards. The 37 Agreement States oversee 19,600 licenses (see Figure 4, page 6). The NRC, Agreement States, and their licensees share a common responsibility to protect public health and safety, security, and the environment.

To address safety and security issues, the NRC has developed regulatory practices, knowledge, and expertise specific to each activity in the nuclear material cycle. Approximately 20 percent of the Nation's electricity is generated by the 104 NRC-licensed commercial nuclear reactors.

Fuel Facilities

The production of nuclear fuel begins at uranium mines where milled uranium ore is used to produce

Figure 3
U.S. COMMERCIAL NUCLEAR POWER REACTORS

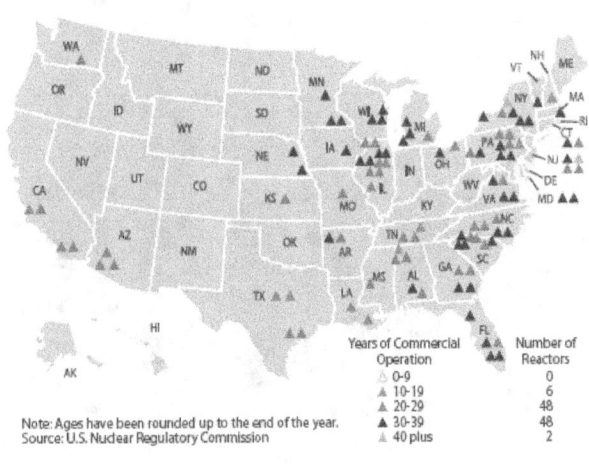

Years of Commercial Operation	Number of Reactors
0-9	0
10-19	6
20-29	48
30-39	48
40 plus	2

Note: Ages have been rounded up to the end of the year.
Source: U.S. Nuclear Regulatory Commission

Figure 4
U.S. MATERIALS LICENSEES

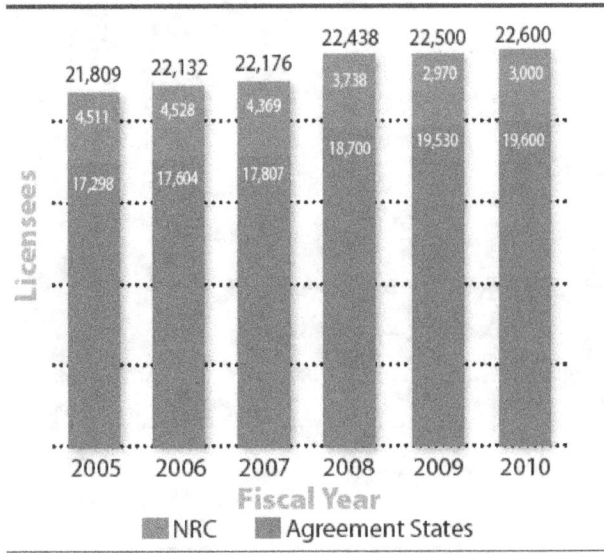

a uranium concentrate called "yellow cake." At a special facility, the yellow cake is converted into uranium hexafluoride gas and loaded into cylinders. The cylinders are sent to a gaseous diffusion plant, where uranium is enriched for use as reactor fuel. The enriched uranium is then converted into oxide powder, fabricated into fuel pellets (each about the size of a fingertip), loaded into metal fuel rods about 3.5 meters long, and bundled into reactor fuel assemblies at a fuel fabrication facility. Assemblies are then transported to nuclear power plants, nonpower research reactor facilities, and naval propulsion reactors for use as fuel. The NRC licenses six operational fuel fabrication and production facilities and three operational enrichment facilities in the United States. Because they handle extremely hazardous material, owners of these facilities take special precautions to prevent theft, diversion by terrorists, and dangerous exposures to workers and the public from this nuclear material.

Reactors

Power plants change one form of energy into another. Electrical generating plants convert heat energy, the kinetic energy of wind or falling water, or solar energy, into electricity. A nuclear power plant converts heat energy into electricity. Other types of heat-conversion plants burn coal, oil, or gas to produce heat energy that is then used to produce electricity. Nuclear energy cannot be seen. There is no burning of fuel in the usual sense. Rather, energy is given off by the nuclear fuel as certain types of atoms split in a process called nuclear fission. This energy is in the form of fast-moving particles and invisible radiation. As the particles and radiation move through the fuel and surrounding water, the energy is converted into heat. The radiation energy can be hazardous, and facilities take special precautions to protect people and the environment from these hazards.

Because the fission reaction produces potentially hazardous radioactive materials, nuclear power plants are equipped with safety systems to protect workers, the public, and the environment. Radioactive materials require careful use because they produce radiation, a form of energy that can damage human cells. Depending on the amount and duration of the exposure, radiation can potentially cause cancer. In a nuclear reactor, most hazardous radioactive substances, called fission byproducts, are trapped in the fuel pellets or in the sealed metal tubes holding the fuel. However, small amounts of these radioactive fission byproducts, principally gases, become mixed with the water passing through the reactor. Other impurities in the water also become radioactive as they pass through the reactor. The facility processes and filters the water to remove these radioactive impurities and then returns the water to the reactor cooling system.

Materials Users

The medical, academic, and industrial fields all use nuclear materials. For example, about one-third of all patients admitted to U.S. hospitals are diagnosed or treated using radioisotopes. Most major hospitals have specific departments dedicated to nuclear medicine. In all, about 112 million nuclear medicine or radiation therapy procedures are performed annually, with the vast majority used in diagnoses. Radioactive

materials used as diagnostic tools can identify the status of a disease and minimize the need for surgery. Radioisotopes give doctors the ability to look inside the body and observe soft tissues and organs, in a manner similar to the way X-rays provide images of bones. Radioisotopes carried in the blood also allow doctors to detect clogged arteries or check the functioning of the circulatory system.

The same property that makes radiation hazardous can also make it useful in treating certain diseases like cancer. When living tissue is exposed to high levels of radiation, cells can be destroyed or damaged. Doctors can selectively expose cancerous cells (cells

Figure 5
SCHEMATIC OF THE NUCLEAR FUEL CYCLE

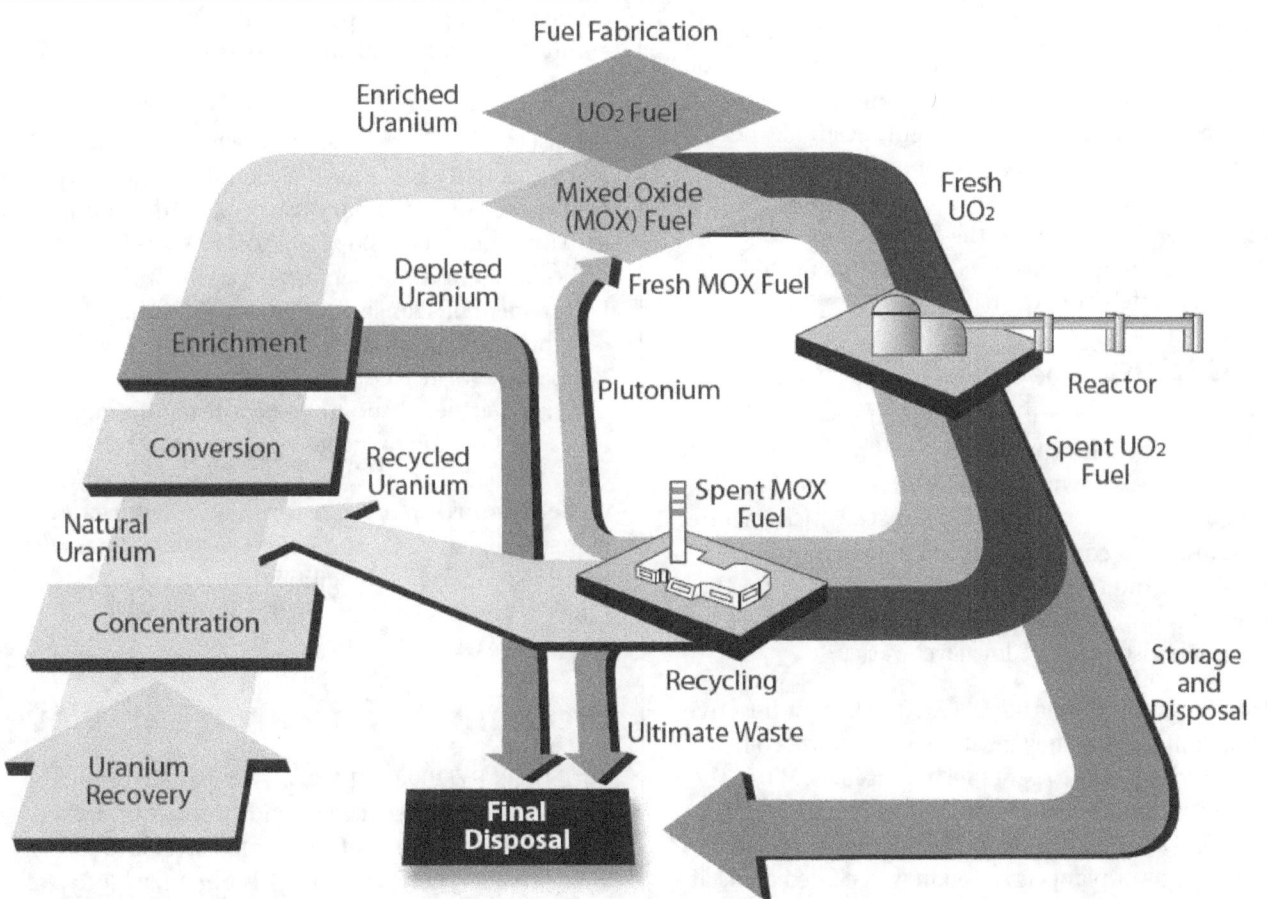

The Mixed Oxide (MOX) fuel is a blend of Plutonium Dioxide and depleted Uranium Dioxide (UO2) that is used as fuel in commercial nuclear power plants.

Source: U.S. Nuclear Regulatory Commission

that are dividing uncontrollably) to radiation to either destroy these cells or damage them so they can no longer reproduce.

Many of today's industrial processes also use nuclear materials. High-tech methods that ensure the quality of manufactured products often rely on radiation generated by radioisotopes. To determine whether a well drilled deep into the ground has the potential for producing oil, geologists use nuclear well-logging, a technique that employs radiation from a radioisotope inside the well to detect the presence of different materials. Radioisotopes are also used to sterilize instruments, find flaws in critical steel parts and welds that go into automobiles and modern buildings, authenticate valuable works of art, and solve crimes by spotting trace elements of poison. Radioisotopes can also eliminate dust from film and compact discs and reduce static electricity (which may create a fire hazard) from can labels. In manufacturing, radiation can change the characteristics of materials, often giving them features that are highly desirable. For example, wood and plastic composites treated with gamma radiation resist abrasion and require low maintenance. As a result, they are used for some flooring in high-traffic areas of department stores, airports, hotels, and churches.

Waste Disposal

During normal operations, a nuclear power plant generates the following two types of radioactive waste: high-level waste, which consists of used fuel (usually called spent fuel), and low-level waste, which includes contaminated equipment, filters, maintenance materials, and resins used in purifying water for the reactor cooling system. Other users of radioactive materials also generate low-level waste.

Nuclear power plants handle each type of radioactive waste differently. They must use special procedures in the handling of the spent fuel because it contains the highly radioactive fission byproducts created while the reactor was operating. Typically, the spent fuel from nuclear power plants is stored in water-filled pools at each reactor site or at a storage facility in Illinois. The water in the spent fuel storage pool provides cooling and adequately shields and protects workers from the radiation. Several nuclear power plants have also

begun using dry casks to store spent fuel. These heavy metal or concrete casks rest on concrete pads adjacent to the reactor facility. The thick layers of concrete and steel in these casks shield workers and the public from radiation.

Currently, most spent fuel in the United States remains stored at individual plants. Permanent disposal of spent fuel from nuclear power plants requires disposal processes and infrastructure that can provide reasonable assurance that the waste will remain isolated for thousands of years. The U.S. Department of Energy (DOE) submitted an application for a permanent spent fuel disposal facility at Yucca Mountain, NV, which was docketed in FY 2008. DOE filed a motion to withdraw its license application with prejudice in FY 2010. The Licensing Board denied DOE's motion. The Commission invited briefing by the parties. The briefing was completed in July 2010, and the case is pending before the Commission.

Licensees often store low-level waste onsite until its radioactivity has diminished and the waste can be disposed of as ordinary trash, or until amounts are large enough for shipment to a low-level waste disposal site in containers approved by the U.S. Department of Transportation. The NRC has developed a waste classification system for low-level radioactive waste based on its potential hazards, and has specified disposal and waste requirements for each of the following general classes of waste: Class A, Class B, and Class C. Generally, Class A waste contains lower concentrations of radioactive material than Class B and Class C wastes. There are two low-level disposal facilities that accept a broad range of low-level wastes, located in Barnwell, SC, and Richland, WA.

Program Performance Overview

The NRC's FY 2008-2013 Strategic Plan determines the agency's long-term goals and strategic direction. The agency has two strategic goals: safety and security. To achieve its goals, the agency is organized into two major programs: the Nuclear Reactor Safety Program, and the Nuclear Materials and Waste Safety Program. The Strategic Plan is located on the NRC Web site at http://www.nrc.gov.

Program Performance Results Scorecard						
Safety Performance Measures	**2005**	**2006**	**2007**	**2008**	**2009**	**2010**
1. Number of new conditions evaluated as red by the Reactor Oversight Process is ≤3.	0	0	0	0	0	0
2. Number of significant accident sequence precursors of a nuclear reactor accident is zero.	0	0	0	0	0	0
3. Number of operating reactors with integrated performance that entered the Manual Chapter 0350 process, the multiple/repetitive degraded cornerstone column, or the unacceptable performance column of the Reactor Oversight Process Action Matrix, with no performance exceeding Abnormal Occurrence Criterion I.D.4, is ≤3.	0	0	1	0	0	0
4. Number of significant adverse trends in industry safety performance, with no trend exceeding Abnormal Occurrence Criterion I.D.4, is ≤1.	0	0	0	0	0	0
5. Number of events with radiation exposures to the public and occupational workers that exceed Abnormal Occurrence Criterion I.A is:						
Reactors: 0.	0	0	0	0	0	0
Materials: ≤2.	1	0	0	0	0	0
Waste: 0.	0	0	0	0	0	0
6. Number of radiological releases to the environment that exceed applicable regulatory limits is:						
Reactor: ≤0.	0	0	0	0	0	0
Materials: ≤2.	0	0	0	0	0	0
Waste: 0.	0	0	0	0	0	0
Security Performance Measures	**2005**	**2006**	**2007**	**2008**	**2009**	**2010**
1. Number of unrecovered losses or thefts of risk-significant radioactive sources is zero.	0	0	0	0	0	0
2. Number of substantiated cases of theft or diversion of licensed, risk-significant radioactive sources or formula quantities of special nuclear material or number of attacks that result in radiological sabotage, is zero.	0	0	0	0	0	0
3. Number of substantiated losses of formula quantities of special nuclear material or substantiated inventory discrepancies of formula quantities of special nuclear material that are caused by theft or diversion or by substantial breakdown of the accountability system is zero.	0	0	0	0	0	0
4. Number of substantial breakdowns of physical security or material control that significantly weaken the protection against theft, diversion, or sabotage is ≤1.	0	0	0	0	0	0
5. Number of significant, unauthorized disclosures of classified and/or safeguards information is zero.	0	0	0	0	0	0

Nuclear Reactor Safety Program

The Nuclear Reactor Safety Program encompasses all NRC efforts to ensure that civilian nuclear power reactor facilities and research and test reactors are licensed and operated in a manner that adequately protects the public health and safety, preserves the environment, and protects against radiological sabotage and theft or diversion of special nuclear materials.

Nuclear Materials and Waste Safety Program

The Nuclear Materials and Waste Safety Program focuses on the safe and secure use of remaining radioactive materials. The Nuclear Materials and Waste Safety Program regulates fuel facilities, medical and industrial nuclear materials users, the disposal of both high-level and low-level waste, the decommissioning of power plants, and the storage and transportation of spent nuclear fuel.

NRC PERFORMANCE MEASURE RESULTS

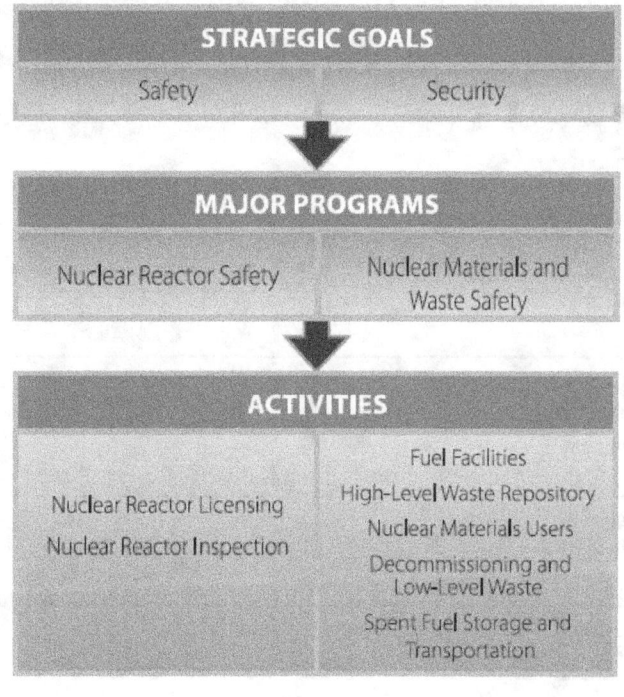

Program Performance Results

STRATEGIC GOAL 1: SAFETY

Ensure Adequate Protection of Public Health and Safety and the Environment

Safety is the primary goal of the NRC. The agency achieves this goal by ensuring that the performance of licensees is at or above acceptable safety levels. NRC safety programs work in conjunction with agency licensees in a partnership. The NRC licensees are responsible for designing, constructing, and operating nuclear facilities safely. The NRC is responsible for regulatory oversight of the licensees. NRC safety goal activities are designed to achieve the strategic outcomes given below.

Strategic Outcomes

- Prevent the occurrence of any nuclear reactor accidents.
- Prevent the occurrence of any inadvertent criticality events.
- Prevent the occurrence of any acute radiation exposures resulting in fatalities.
- Prevent the occurrence of any releases of radioactive materials that result in significant radiation exposures.
- Prevent the occurrence of any releases of radioactive materials that cause significant adverse environmental impacts.

FY 2010 Results

In FY 2010, the NRC achieved all five of its safety goal strategic outcomes. The NRC also uses six performance measures to determine whether it has met its safety goal. The agency met all six performance measure targets in FY 2010.

The first three performance measures focus on performance at individual nuclear power plants. Inspection results show that all of the nuclear power plants are operating safely. The fourth measure tracks the trends of several key indicators of nuclear power plant safety. This measure is the broadest measure of the safety of nuclear power plants, incorporating the performance results from all plants to determine industry average results. It shows that there were no statistically significant adverse trends in any of the indicators in FY 2010.

The last two safety performance measures track harmful radiation exposures to the public and occupational workers, and radiation exposures that harm the environment. There were no harmful human or environmental exposures in FY 2010.

Safety Goal Strategies

The agency uses the following strategies to guide its activities and achieve its safety goal:

(1) Develop, implement, and maintain licensing and regulatory programs for reactors, fuel facilities, materials users, spent fuel management, uranium recovery, and decommissioning activities to ensure the adequate protection of public health, safety, and the environment.

(2) Continue to oversee the safe operation of existing power plants while preparing for and managing the review of applications for new power reactors.

(3) Conduct NRC safety, security, and emergency preparedness programs.

(4) Improve the NRC's regulatory programs and apply safety-focused research to anticipate and resolve safety issues.

(5) Use sound science and state-of-the-art methods to establish, where appropriate, risk informed and performance-based regulations.

(6) Promote attention to safety matters and individual accountability for those engaged in regulated activities.

(7) Use domestic and international operating experience to inform decisionmaking.

(8) Oversee licensee safety performance through inspections, investigations, enforcement, and performance assessment activities.

(9) Effectively respond to events at NRC licensed facilities and other events of national interest, including maintaining and enhancing the NRC's critical incident response and communication capabilities.

STRATEGIC GOAL 2: SECURITY

Ensure Adequate Protection in the Secure Use and Management of Radioactive Materials

The NRC must remain vigilant in ensuring the security of nuclear facilities and materials in an elevated threat environment. The agency achieves its common defense and security goal using licensing and oversight programs similar to those employed in achieving its safety goal. NRC's security activities are designed to achieve the strategic outcome given below.

Strategic Outcome

■ Prevent any instances where licensed radioactive materials are used domestically in a manner hostile to the security of the United States.

FY 2010 Results

In FY 2010, the NRC achieved its security goal strategic outcome. The NRC also uses five security performance measures to determine whether the agency has met its security goal. The agency met all five performance measure targets in FY 2010. The first performance measure tracks unrecovered losses or thefts of risk-significant radioactive sources. The measure ensures that those radioactive sources that the agency has determined to be risk-significant to the public health and safety are accounted for at all times. The ability to account for these sources is critical to secure the Nation from "dirty bomb" attacks or other means of radiation dispersal.

The second, third, and fourth performance measures evaluate the number of significant security events and incidents that occur at NRC-licensed facilities. These measures determine whether nuclear facilities maintain adequate protective forces to prevent theft or diversion of nuclear material or sabotage; whether systems in place at licensee plants accurately account for the type and amount of materials processed, used, or stored and whether the facilities account for special nuclear material at all times with no losses of this material. No events met the conditions for any of these measures in FY 2010.

The last security measure tracks significant unauthorized disclosures of classified or safeguards information that may cause damage to national security or public safety. This measure focuses on whether classified information or safeguards information is stored and used in such a way as to prevent its disclosure to terrorist organizations, other nations, personnel without a need to know, or the public. Unauthorized disclosures can harm national security or compromise public health and safety. This measure also focuses on whether controls are in place to maintain and secure the various devices and systems (electronic or paper-based) that the agency and its licensees use to store, transmit, and use this information. There were no documented disclosures of this type of information during FY 2010.

Security Goal Strategies

The agency uses the following strategies to guide its activities and achieve its security goal:

(1) Use relevant intelligence information and security assessments to maintain realistic and effective security requirements and mitigation measures.

(2) Share security information with appropriate stakeholders and international partners.

(3) Oversee licensee security performance through inspections and force-on-force exercises.

(4) Control the handling and storage of sensitive security information, and the communication of information to licensees and Federal, State, and local partners.

(5) Support Federal response plans that employ an approach to the security of nuclear facilities and radioactive material that integrates the efforts of licensees and Federal, State, local, and Tribal governments.

(6) Use a risk informed approach to implement appropriate regulatory controls for the possession, handling, import, export, and transshipment of radioactive materials.

(7) Enhance the programs for control of the security of radioactive sources and strategic special nuclear material commensurate with their risk, including enhancements required by the Energy Policy Act of 2005.

(8) Promote U.S. national security interests and nuclear nonproliferation policy objectives for NRC-licensed imports and exports of source and special nuclear materials and nuclear equipment.

Data Completeness and Reliability

The NRC considers the data contained in this report to be complete, reliable, and relevant. The data are complete because the agency reports actual performance data for every performance goal and indicator in the report. The agency considers the data in this report to be reliable and relevant, because they have been validated and verified. A report entitled, "Verification and Validation of the NRC's Measures and Metrics" is available on the NRC's Web site at http://www.nrc.gov.

Future Challenges

The NRC ensures that the health and safety of the American public and the environment are adequately protected from any harmful effects of using nuclear materials. The nuclear industry has experienced a substantial improvement in safety at nuclear power plants over the past 35 years as both the nuclear industry and the NRC have gained substantial experience in the operation and maintenance of nuclear power facilities. Despite this excellent safety and security record, the agency cannot rest on its achievements.

The primary challenges the NRC faces are the large number of new nuclear plants that have applied for licenses, the safe disposal of high-level nuclear waste, and the need to ensure security at nuclear facilities.

New Nuclear Power Plants

With increased concerns about the continued availability and cost of oil as well as concerns over the environmental damage caused by coal-burning electrical plants, the amount of electricity supplied by nuclear power is likely to increase substantially in the future. The NRC last issued a nuclear power plant construction permit in 1977. Since 2007, the agency has received 18 Combined Operating License (COL) applications for sites across the country. The agency's primary challenge is to license new reactors to ensure that they will operate safely as they provide electricity required by the Nation for economic growth. However, before licensing any new nuclear reactor, the agency requires a detailed analysis of new reactor designs. This analysis includes a study of the reactor's vulnerability to accidents and security compromises. It also includes the development of inspection procedures, tests, analyses, and acceptable criteria for construction. The NRC also evaluates commercial gas centrifuge facilities that use new methods of enriching nuclear fuel for reactors.

Safe Disposal of High-Level Waste

Safely disposing of the waste from nuclear power plants is vital to protecting public health and the environment. In FY 2008, DOE filed a license application to establish the Nation's first repository for high-level radioactive waste at Yucca Mountain, NV. The NRC staff accepted and docketed the application. On March 3, 2010, DOE filed a motion seeking to withdraw its license application, with prejudice. On June 29, 2010, the Licensing Board denied DOE's motion. The Commission invited briefing by the parties. The briefing was completed in June 2010 and the case is pending before the Commission. The NRC continued to conduct a technical review of the application during FY 2010 and published the first volume of the Yucca Mountain Safety Evaluation Report.

Most nuclear waste is now safely and securely stored at reactor sites. In addition to the storage of nuclear waste, safely transporting spent nuclear fuel is a significant issue for the public and the agency. More than 1,300 spent fuel shipments regulated by the NRC have been safely transported in the United States in the past 25 years. The agency must be able to assure the public that all movements of nuclear waste will be safe and secure.

Security at Nuclear Facilities

In addition to safety, the security of nuclear materials is of paramount importance to the Nation. Nuclear facilities are among the most secure facilities in the United States. The NRC, in concert with other Federal agencies, constantly monitors intelligence to determine the level of threat faced by nuclear facilities. The agency continues to improve regulatory requirements to better ensure the security of nuclear materials and facilities. The threat faced by the Nation from those seeking to steal classified information has become more urgent in recent years. Nuclear facilities have implemented more and enhanced security measures, including "force-on-force" training exercises, to help ensure protection of this vital national infrastructure.

The NRC is collaborating with both the Federal Energy Regulatory Commission and the North American Electric Reliability Corporation (NERC) to ensure that nuclear safety and security are maintained at nuclear facilities while trying to optimize Bulk Power System reliability. The NRC has also implemented a process to inform licensees of emergent cyber security issues by posting Department of Homeland Security cyber security bulletins, alerts, reports, and advisories to its protected Web server.

Financial Performance Overview

As of September 30, 2010, the financial condition of the NRC was sound with respect to having sufficient funds to meet program needs and adequate control of these funds in place to ensure obligations did not exceed budget authority. The NRC prepared its financial statements in accordance with the accounting standards codified in the Statements of Federal Financial Accounting Standards (SFFAS) and Office of Management and Budget (OMB) Circular A-136, "Financial Reporting Requirements."

Sources of Funds

The NRC has two appropriations: Salaries and Expenses and the Office of the Inspector General. Funds for both appropriations are available until expended. The NRC's total new FY 2010 budget authority was $1,066.9 million. Of this amount, $1,056.0 million was for the Salaries and Expenses appropriation and $10.9 million was for the Office of the Inspector General appropriation. This represents an increase in new budget authority of $21.4 million over FY 2009 ($21.4 million for the Salaries and Expenses appropriation, including a decrease of $20.0 million for the Nuclear Waste Fund, and no change for the Office of the Inspector General appropriation). In addition, $76.0 million from prior-year appropriations (net of the $18.0 million rescission of prior year funds), $9.6 million from prior-year reimbursable work, and $11.3 million for new reimbursable work to be performed for others was available to obligate in FY 2010. The sum of all funds available to obligate for FY 2010 was $1,163.8 million, which was a $1.4 million decrease from the FY 2009 amount of $1,165.2 million (see Figure 6).

Figure 6
SOURCES OF FUNDS (Projected)

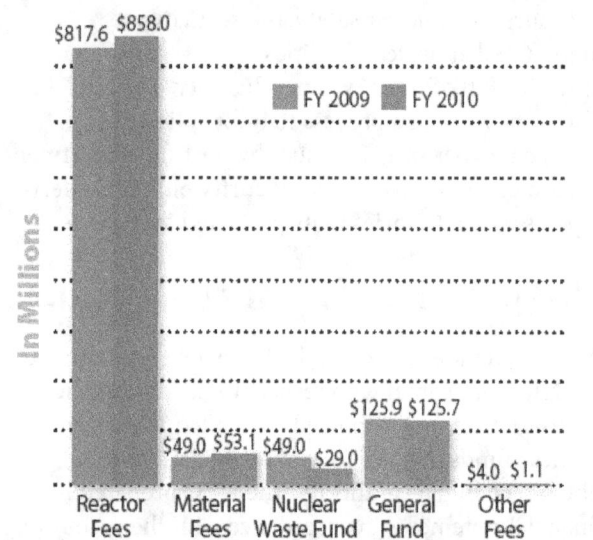

The Omnibus Budget Reconciliation Act of 1990 (OBRA-90), as amended, requires the NRC to collect fees to offset approximately 90 percent of its new budget authority, less the amount appropriated to the NRC from the Nuclear Waste Fund, and amounts appropriated for waste incidental to reprocessing and generic homeland security for FY 2010. The projected amount to be received from reactor and material fees in FY 2010 was $911.1 million after accounting for billing adjustments. The NRC collected $909.5 million of the required $912.2 million in fees for the year which was 99.7 percent of the 90 percent fee recovery requirement.

Uses of Funds by Function

The NRC incurred obligations of $1,119.1 million in FY 2010, which was an increase of $35.0 million over FY 2009 (see Figure 7). Approximately 54 percent of obligations were used for salaries and benefits. The remaining 46 percent was used to obtain technical assistance for the NRC's principal regulatory programs, to conduct confirmatory safety research, and to cover operating expenses (e.g., building rentals, transportation, printing, security services, supplies, office automation, training), staff travel, and reimbursable work. The unobligated budget authority available at the end of FY 2010 was $44.7 million, a $36.4 million decrease compared to the FY 2009 amount of $81.1 million. Of the $44.7 million, $10.9 million is for reimbursable work and $33.8 million is available to fund critical NRC needs in FY 2011.

Figure 7
USES OF FUNDS BY FUNCTION

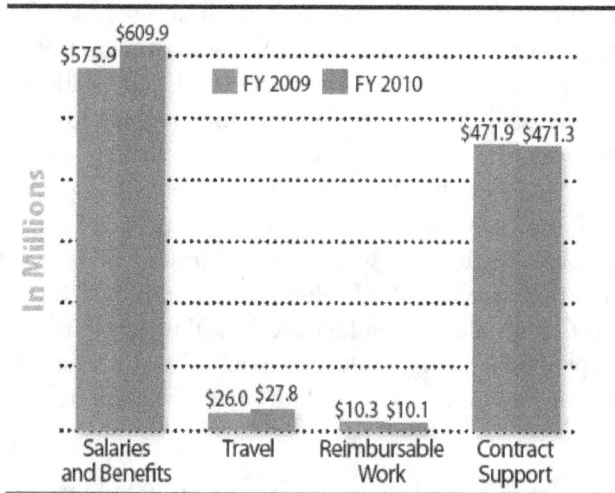

Audit Results

The NRC received an unqualified audit opinion on its FY 2010 financial statements and internal controls. The auditors found no instances of noncompliance or substantial noncompliance with laws and regulations during the FY 2010 audit.

A summary of the Financial Statement Audit Results is included in the "Other Accompanying Information" section of this report.

Limitations of the Financial Statements

The principal financial statements have been prepared to report the financial position and results of operations of the NRC, pursuant to the requirements of 31 U.S.C. 3515(b). While the statements have been prepared from books and records of the NRC in accordance with Generally Accepted Accounting Principles (GAAP) for Federal entities and with the formats prescribed by the Office of Management and Budget (OMB), the statements are in addition to the financial reports used to monitor and control budgetary resources, which are prepared from the same books and records. The statements should be read with the understanding that they are for a component of the U.S. Government, a sovereign entity.

Financial Statement Highlights

The NRC's financial statements summarize the financial activity and financial position of the agency. The financial statements, footnotes, and required supplementary information, appear in Chapter 3, "Financial Statements and Auditors' Report." Analysis of the principal statements follows.

Analysis of the Balance Sheet

Assets. The NRC's assets were $590.3 million as of September 30, 2010, a decrease of $21.5 million from the end of FY 2009. The decrease is primarily due to the Fund Balance with Treasury decreasing by $28.5 million.

ASSET SUMMARY (In Millions)

As of September 30,	2010	2009
Fund Balance with Treasury	$420.1	$448.6
Accounts Receivable, Net	130.9	128.2
Property & Equipment, Net	36.2	31.6
Other	3.1	3.4
Total Assets	$590.3	$611.8

The Fund Balance with Treasury was $420.1 million at September 30, 2010, accounting for 71 percent of total assets. This account represents appropriated funds, collected license fees, and other funds maintained at the U.S. Department of Treasury (Treasury) to pay current liabilities and to finance authorized purchase commitments. The $28.5 million decrease in the fund balance is primarily the result of increases of $47.3 million in general disbursements, $30.9 million in salaries and benefits, and $11.4 million in grant disbursements which decreased the fund balance; offset by a $55.2 million beginning balance increase over the prior year. The fund balance had a net increase of $3.4 million resulting from an increase in appropriated funds of $21.4 million over FY 2009 as a result of new budget authority (including a decrease of $20.0 million for the Nuclear Waste Fund) reduced by a $18.0 million rescission of prior year unobligated funds returned to Treasury. During the year, fees collected, and then transferred to Treasury, increased $51.7 million over FY 2009 having a net offsetting effect on the fund balance. The revenue generated by fees assessed to licensees as required by law is sent to Treasury to offset approximately 90 percent of the NRC's appropriations received during the year.

Accounts receivable consists of amounts owed to the NRC by other Federal agencies and the public. Accounts Receivable, Net, as of September 30, 2010, was $130.9 million, which includes an offsetting allowance for doubtful accounts of $2.9 million. The 2 percent increase from the FY 2009 year-end Accounts Receivable, Net, balance of $128.2 million is primarily due to intragovernmental fee receivables and reimbursements.

LIABILITIES SUMMARY (In Millions)

As of September 30,	2010	2009
Accounts Payable	$ 40.5	$ 51.0
Federal Employee Benefits	7.6	7.6
Other Liabilities	112.0	86.2
Total Liabilities	$160.1	$144.8

Liabilities. Total liabilities were $160.1 million as of September 30, 2010, an increase of $15.3 million from the FY 2009 year-end balance of $144.8 million. The change in Total Liabilities is due to an increase in Other Liabilities of $25.8 million, which is comprised of a new contingent liability recorded in FY 2010 of $11.8 million for the probable likelihood of an adverse outcome of legal claims, and increases over FY 2009 of $6.8 million for grants payable due to a rise in the number and dollar volume of the NRC's grant programs, $3.1 million in accrued annual leave, and $3.5 million in accrued funded salaries and benefits. This was offset by a decrease in Accounts Payable of $10.5 million due to a decrease of the accounts payable accrual and early payment of invoices scheduled to be paid in the first month of FY 2011 to prepare for the implementation of the new integrated financial management system, which was effective at the beginning of FY 2011.

Of the agency's liabilities, $71.5 million were not covered by budgetary resources, a 26 percent increase over the balance of $56.6 million as of September 30, 2009. The increase of $14.9 million is primarily due to the contingent liability in FY 2010 of $11.8 million and an increase in unfunded accrued annual leave of $3.1 million. The liabilities not covered by budgetary resources at September 30, 2010 include $50.4 million in unfunded accrued annual leave for the amount of leave earned but not yet taken, $11.8 million for contingent liabilities and $9.3 million in accrued and future workers' compensation.

NET POSITION SUMMARY (In Millions)

As of September 30,	2010	2009
Unexpended Appropriations	$311.9	$338.6
Cumulative Results of Operations	118.3	128.4
Total Net Position	$430.2	$467.0

Net Position. Total Net Position, which is the difference between Total Assets and Total Liabilities, was $430.2 million as of September 30, 2010, a decrease of $36.8 million from the FY 2009 year-end balance. Net Position is comprised of two components: Unexpended Appropriations and Cumulative Results of Operations. Unexpended Appropriations is the amount of spending authority granted by Congress that remains unused by the agency. The decrease in FY 2010 for Unexpended Appropriations is $26.7 million. Cumulative Results of Operations which represents the cumulative excess of financing sources over expenses, decreased $10.1 million.

Analysis of the Statement of Net Cost

Net costs are gross costs offset by earned revenue. The Statement of Net Cost presents the net cost of the NRC's two programs as identified in the NRC Annual Performance Plan. The purpose of this statement is to link program performance to the cost of programs. The NRC's Net Cost of Operations for the year ended September 30, 2010, was $217.0 million, which is an increase of $46.6 million over the FY 2009 net cost of $170.4 million.

NET COST OF OPERATIONS (In Millions)

For the years ended September 30,	2010	2009
Nuclear Reactor Safety and Security	$ 46.3	$ 2.9
Nuclear Materials & Waste Safety and Security	170.7	167.5
Net Cost of Operations	$217.0	$170.4

NRC's total gross costs increased $97.6 million. Gross costs increased $85.7 million in Nuclear Reactor Safety and Security primarily due to increases of $24.0 million in salaries and benefits and $70.2 million in contractor support. These increases were primarily for new reactor activities, existing licensing and oversight activities, and international activities. The Nuclear Materials & Waste Safety and Security program gross costs increased $11.9 million primarily due to increases in activities for nuclear materials licensing, fuel facilities, and spent fuel storage and transportation; offset by a decrease in costs for high level waste activities.

Total earned revenue increased $51.0 million from $872.5 million for the year ended September 30, 2009, to $923.5 million on September 30, 2010. Earned revenue increased for the Nuclear Reactor Safety and Security program by $42.3 million and for the Nuclear Materials & Waste Safety and Security program by $8.7 million. The increases are primarily the result of increases in fees collected due to the increase in appropriations for NRC activities, of which the NRC is required to collect approximately 90 percent through fee billing. Fees for reactor and materials licensing and inspections are collected in accordance with Title 10 of the *Code of Federal Regulations* (10 CFR) Part 170, "Fees for Facilities, Materials, Import and Export Licenses, and Other Regulatory Services under the Atomic Energy Act of 1954, as Amended," and 10 CFR Part 171, "Annual Fees for Reactor Licenses and Fuel Cycle Licenses and Materials Licenses, Including Holders of Certificates of Compliance, Registrations, and Quality Assurance Program Approvals and Government Agencies Licensed by the NRC."

Analysis of the Statement of Changes in Net Position

The Statement of Changes in Net Position reports the change in net position during the reporting period. Net position is affected by changes in its two components—Cumulative Results of Operations and Unexpended Appropriations. The decrease in Net Position of $36.8 million from FY 2009 to FY 2010 is due to decreases of $10.1 million in Cumulative Results of Operations and $26.7 million in Unexpended Appropriations.

A change in Cumulative Results of Operations results from changes in the beginning balance, financing sources, and the net cost of operations. The decrease of $10.1 million is primarily due to the change of $46.5 million in the net cost of operations exceeding the increase in financing sources of $36.4 million. The financing sources primarily included increases of $47.8 million in appropriations used and $8.6 million in imputed financing from costs absorbed by others including imputed costs for retirement and health

benefits; offset by a decrease in the Nuclear Waste Fund transfer of $20.0 million.

A change in unexpended appropriations primarily results from appropriations received and adjustments (rescissions, etc.) being more, or less, than appropriations used during the fiscal year. In FY 2010, appropriations received of $128.4 million consisted of NRC's total appropriation of $1,066.9 million, reduced by $909.5 million in fee collections returned to Treasury and $29.0 million for the Nuclear Waste Fund transfer. A rescission of $18.0 million of prior year unobligated funds reduced unexpended appropriations. Appropriations used in FY 2010 totaled $137.1 million and consisted of funds used of $1,079.7 million reduced by collection from fees assessed of $909.5 million and Nuclear Waste Fund expenses of $33.1 million.

Analysis of the Statement of Budgetary Resources

The Statement of Budgetary Resources reports the source and status of budgetary resources at the end of the period. It presents the relationship between budget authority and budget outlays, and the reconciliation of obligations to total outlays. For FY 2010, the NRC had total budgetary resources available of $1,163.8 million which remained at basically the same level as FY 2009 at $1,165.2 million. During the year, budgetary resources decreased by $18.0 million for a rescission of prior year funds which were returned to Treasury in FY 2010. The appropriation received during FY 2010 increased $21.4 million, from $1,045.5 million in FY 2009 to $1,066.9 million increasing budgetary resources. The appropriation included increases of $20.5 million for the Nuclear Reactor Safety and Security program, $0.9 million for the Nuclear Materials and Waste Safety and Security program, and no change for the Office of the Inspector General. This funding provided for increases in contract support services primarily for new and existing reactor activities and regulatory oversight of existing reactors.

For FY 2010, the NRC had Obligations Incurred of $1,119.1 million, compared to FY 2009 Obligations Incurred of $1,084.1 million, an increase of $35.0 million. The increase is due primarily to an increase of $56.9 million in obligations for NRC disbursements, offset by a decrease of $21.9 million in obligations relating to the Nuclear Waste Fund. Obligations Incurred also includes reimbursable obligations which remained at the same level as the prior year.

Gross outlays for FY 2010 were $1,088.7 million, which represents an $89.6 million increase from FY 2009 total outlays of $999.1 million. The increase is due to an increase in general disbursements of $47.3 million, salary and benefits disbursements of $30.9 million, and grant disbursements of $11.4 million. Gross outlay increases of $79.1 million in the Nuclear Reactor Safety and Security program, primarily reflected new reactor and existing reactor licensing activities, and existing reactor oversight. Gross outlays for the Nuclear Materials and Waste Safety and Security program showed an increase of $10.5 million primarily due to increased outlays for activities related to materials licensing, fuel facilities, and spent fuel storage and transportation; offset by decreased outlays for the high level waste program.

Systems, Controls, and Legal Compliance

Management Assurances

This section provides information on NRC's compliance with the *Federal Managers' Financial Integrity Act of 1982* (the Public Law 97-255), (Integrity Act) OMB Circular A-123,

Management's Responsibility for Internal Control, and the Federal Financial Management Improvement Act of 1996.

Federal Managers' Financial Integrity Act

The Federal Managers' Financial Integrity Act (Integrity Act of 1982) mandates that agencies establish internal control to provide reasonable assurance that the agency: complies with applicable laws and regulations; safeguards assets against waste, loss, unauthorized use, or misappropriation; and properly accounts for and records revenues and expenditures. The Integrity Act encompasses program, operational, and administrative

U.S. NUCLEAR REGULATORY COMMISSION

FISCAL YEAR 2010

FEDERAL MANAGERS' FINANCIAL INTEGRITY ACT STATEMENT

The U.S. Nuclear Regulatory Commission (NRC) managers are responsible for establishing and maintaining effective internal control and financial management systems that meet the objectives of the Federal Managers' Financial Integrity Act (Integrity Act). The NRC conducted its assessment of internal control over the effectiveness and efficiency of operations and compliance with applicable laws and regulations, and in accordance with OMB Circular A-123, Management's Responsibility for Internal Control. Based on the results of this evaluation, the NRC can provide reasonable assurance that its internal control over the effectiveness and efficiency of operations and compliance with applicable laws and regulations as of September 30, 2010, was operating effectively and no material weaknesses were found in the design or operation of internal control.

In addition, NRC conducted its assessment of the effectiveness of internal control over financial reporting, which includes safeguarding of assets and compliance with applicable laws and regulations, in accordance with the requirements of Appendix A of OMB Circular A-123. Based on the results of the evaluation, NRC can provide reasonable assurance that NRC's internal control over financial reporting as of June 30, 2010, was operating effectively, and no material weaknesses were found in the design or operation of the internal control over financial reporting.

The NRC can also provide reasonable assurance that its financial systems comply with the requirements of the Integrity Act and with the component requirements of the Federal Financial Management Improvement Act.

Gregory B. Jaczko
Chairman
U.S. Nuclear Regulatory Commission
November 1, 2010

areas, as well as accounting and financial management. It also requires the Chairman to provide an assurance statement on the adequacy of internal controls and on the conformance of financial systems to Government-wide standards.

Internal Control Program

Internal controls are the organization, policies, and procedures to help program and financial managers achieve results and safeguard the integrity of their programs. NRC managers are responsible for designing and implementing effective internal controls in their areas of responsibility. Each office director and regional administrator prepares an annual assurance certification that identifies any control weaknesses requiring the attention of the NRC Executive Committee on Internal Control (ECIC). These certifications are based on internal control activities such as risk assessments, and on other activities such as program evaluations, management reviews, self-assessments, senior leadership meetings, agency lessons learned review board meetings, agency action review meetings, audits of financial statements, reviews of financial statements, Inspector General and U.S. Government Accountability Office audits and reports, and other information provided by the congressional committees of jurisdiction.

The ECIC consists of senior executives from the Office of the Chief Financial Officer and the Office of the Executive Director for Operations. The agency's General Counsel and Inspector General participate as advisors.

The ECIC met and reviewed the reasonable assurance certifications provided by the offices and regions. The ECIC then informed the Chairman as to whether the NRC had any internal control deficiencies serious enough to require reporting as a weakness or noncompliance.

The NRC's internal control program requires that internal control deficiencies be documented and reported in office and regional internal control plans and operating plans. The internal control plans provide for annual updates, and the operating plan process provides for quarterly updates. Both ensure that key issues receive senior management attention.

Combined with the individual assurance statements discussed previously, the internal control information in these plans provides the framework for monitoring and improving the agency's internal control on an ongoing basis.

FY 2010 Integrity Act Results

The NRC evaluated its internal control systems for the fiscal year ending September 30, 2010. Based on this evaluation, the NRC is able to provide a statement of assurance that the internal controls and financial systems meet the objectives of the Integrity Act. The NRC has reasonable assurance that its internal controls are effective and that its financial management systems conform to Governmentwide standards.

Office of Management and Budget Circular A-123, "Management's Responsibility for Internal Control," including Appendix A, "Internal Control over Financial Reporting"

In FY 2006, the NRC implemented the requirements of the revised OMB Circular A-123, which defined and strengthened management's responsibility for internal control in Federal agencies. The revised circular included updated internal control standards. Appendix A requires Federal agencies to assess the effectiveness of internal controls over financial reporting and to prepare a separate annual statement of assurance as of June 30, 2010.

In FY 2007, the NRC adopted a 3-year rotational testing plan. The agency determined that three of the original nine key processes were significant enough to include in the testing each year of the 3-year cycle. The remaining six key processes were to be tested once in the 3-year cycle, two each year. In FY 2010, the NRC continued its assessment of internal control over financial reporting. The agency reevaluated its scope of financial reports, materiality values, risk assessments, key processes, and key controls. Based on the results of this evaluation, the NRC can provide reasonable assurance that its internal control over financial reporting was operating effectively as of June 30, 2010, and that the evaluation found no material weakness in the design or operation of the internal controls over financial reporting.

Federal Financial Management Improvement Act

The Improvement Act of 1996 requires each agency to implement and maintain systems that comply substantially with (1) Federal financial management system requirements, (2) applicable Federal accounting standards, and (3) the standard general ledger at the transaction level. The Improvement Act requires the Chairman to determine whether the agency's financial management systems comply with the Improvement Act and to develop remediation plans for systems that do not comply.

Financial Management Systems Strategies

The NRC has started a business transformation initiative to develop an enterprise-wide financial system. The NRC plans to complete our business transformation in four distinct phases (or implementations). The four phases will cover the agency's core financial, acquisition, time and labor and budget formulation functions respectively. The objective is to consolidate and automate data and processes within a single business solution to make the NRC a more transparent, efficient and effective organization.

During FY 2010, the first phase of our transformation was completed and five stand-alone legacy core financial systems were consolidated with nine subsystems into a new commercial-off-the-shelf core financial system (CFS). In FY 2013, the second phase of our transformation will be completed by integrating the agency's acquisition function with the CFS. After FY 2013 the plan is to complete our objective for an integrated and consolidated enterprise financial and acquisitions management system by consolidating the Agency's time and labor and budget formulation functions with the core financial and acquisitions functions within a single business solution.

FY 2010 Improvement Act Results

As of September 30, 2010, the NRC evaluated its financial systems to determine whether they complied with applicable Federal requirements and accounting standards required by the Improvement Act. The NRC evaluated eight systems: e-Travel, Federal Financial System, Federal Personnel Payroll System, Human Resources Management System, Cost Accounting System, Capitalized Property System, Fee Billing System, and Budget Formulation System. As of September 30, 2010, the agency's financial management systems were in compliance with the Improvement Act. In making this determination, the NRC considered all available information, including the report from the ECIC on the effectiveness of internal controls, the Office of the Inspector General audit reports, and the results of the agency's financial management system reviews. The agency also relied on the U.S. Department of the Interior National Business Center (DOI-NBC) annual reasonable assurance statement, which concluded that, for FY 2010, the cross-serviced financial systems were in substantial compliance with Federal financial management system requirements.

Prompt Payment

The Prompt Payment Act of 1982, as amended, requires Federal agencies to make timely payments to vendors for supplies and services, to pay interest penalties when payments are made after the due date, and to take cash discounts when they are economically justified. In FY 2010, the NRC paid 98 percent of the 13,372 invoices subject to the Prompt Payment Act on time (see Figure 8). The NRC incurred $3,143 in interest penalties during FY 2010.

Figure 8
PROMPT PAYMENT

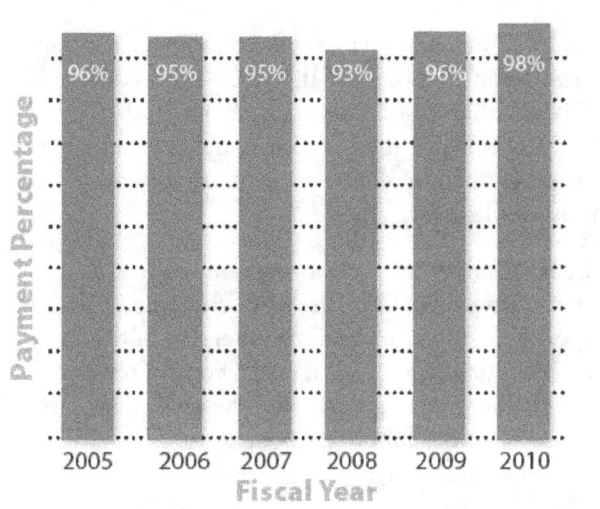

Improper Payments

The NRC remains at low risk of making improper payments. At the present time, the NRC's payments consist of commercial vendor, interagency, and travel reimbursements. The NRC monitors and reports improper payments within its programs and continues to evaluate internal controls guarding against improper payments. The NRC continues to perform annual risk assessments for each of these areas. Based on the FY 2010 risk assessments, the number and amount of improper payments fall below the external reporting requirement established by OMB guidance on what is considered a significant risk. The NRC awards less than $500 million in annual contracts and, therefore, is not subject to annual reporting under the Recovery Auditing Act. The DOI-NBC's Federal Personnel/Payroll System, as the system of record for payroll disbursements, is responsible for monitoring and reporting on any improper payroll-related payments.

Debt Collection

The Debt Collection Improvement Act of 1996 enhances the ability of the Federal Government to service and collect debts. The agency's goal is to maintain the level of delinquent debt owed to the NRC at year end to less than 1 percent of its annual billings. The NRC continues to meet this goal and, at the end of FY 2010 delinquent debt was $2.5 million (Figure 9). The NRC continues to pursue the collection of delinquent debt and refers all eligible debt over 180 days delinquent to the U.S. Department of the Treasury for collection.

Figure 9
DELINQUENT DEBT

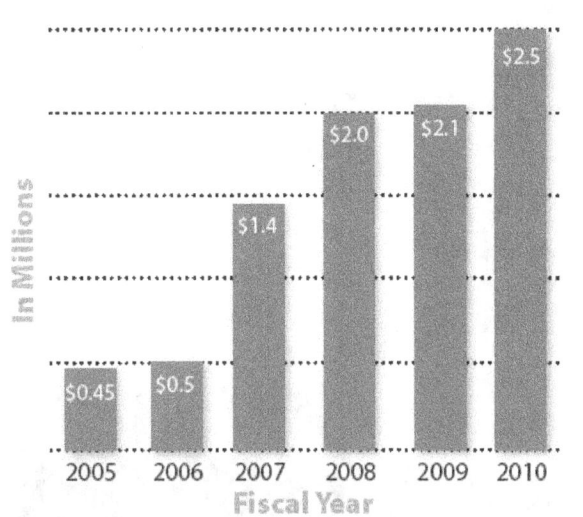

Biennial Review of User Fees

The Chief Financial Officers Act requires agencies to conduct a biennial review of fees, royalties, rents, and other charges imposed by agencies, and make revisions to cover program and administrative costs incurred. Each year, the NRC revises the hourly rates for license and inspection fees and adjusts the annual fees to meet the fee collection requirements of the Omnibus Budget Reconciliation Act of 1990, as amended. The most recent changes to the license, inspection, and annual fees are described in the *Federal Register* (75 FR 34219, June 16, 2010).

The fees and charges for the Criminal History Program and the Freedom of Information Act requests were also revised to more appropriately recognize actual costs. No other reviews were completed this year.

Inspector General Act of 1978

The NRC has established and continues to maintain an excellent record in resolving and implementing Office of the Inspector General open audit recommendations. In the "Other Accompanying Information" section of this report, "Management Decisions and Final Actions on OIG Audit Recommendations," includes this information, as well as data concerning disallowed costs determined through contract audits conducted by the Defense Contract Audit Agency.

Photo Courtesy of NRC Photo Library

The NRC Chairman, Commissioners, and the Executive Director for Operations with the Keynote Speaker for the NRC All-Hands Meeting, Dr. Roger Dean Duncan - September 29, 2010.

Chapter 2

Program Performance

Photo Courtesy of NRC Photo Library

Point Beach Nuclear Power Plant, Two Creeks, WI.

Photo Courtesy of NRC Photo Library

Decommissioned reactor vessel on a transporter.

Measuring and Reporting Performance

This chapter presents information on the U.S. Nuclear Regulatory Commission's (NRC's) performance in achieving its mission during fiscal year (FY) 2010. The agency's mission is to license and regulate the Nation's civilian use of byproduct, source, and special nuclear materials to ensure adequate protection of public health and safety, promote the common defense and security, and protect the environment.

This chapter describes the NRC's performance results and program achievements in accomplishing its two strategic goals of safety and security. The Safety goal section addresses the NRC's activities that regulate operating reactors, new reactors, fuel facilities, nuclear material users, decommissioning and low-level waste, spent fuel storage and transportation, and the proposed high-level waste repository. The Security goal section addresses the agency's security activities. In addition, this chapter describes the agency's progress in achieving its organizational excellence objectives of openness, effectiveness, and operational excellence. Finally, it describes information on data sources, data quality, and the completeness and reliability of performance data. The discussion focuses primarily on the NRC's methods for collecting and analyzing data and ensuring data security.

Goals and Performance Measures

STRATEGIC GOAL 1: SAFETY
Ensure Adequate Protection of Public Health and Safety and the Environment

Strategic Outcomes

The NRC's strategic outcomes specify those outcomes that correlate with the NRC meeting its Safety goal. The NRC's Safety goal has five strategic outcomes that must occur for the agency to achieve its objective to ensure adequate protection of public health and safety and the environment:

- Prevent the occurrence of any nuclear reactor accidents.
- Prevent the occurrence of any inadvertent criticality events.
- Prevent the occurrence of any acute radiation exposures resulting in fatalities.
- Prevent the occurrence of any releases of radioactive materials that result in significant radiation exposures.
- Prevent the occurrence of any releases of radioactive materials that cause significant adverse environmental impacts.

RESULTS: In FY 2010, the NRC met all of the agency's Safety goal strategic outcomes.

Performance Measures

The NRC uses performance measures to assess whether the agency has met its Safety goal. Performance measures are set at a different risk level than the strategic outcomes, and missing a performance measure signals that safety levels may have deteriorated at the agency strategic planning level. If the NRC misses a performance measure, the agency will take corrective actions to bring the measure back into the target range. Table 1 shows the agency's annual performance measures and their outcomes for the past 6 years.

Table 1
SAFETY GOAL PERFORMANCE MEASURES

Performance Measure	2005	2006	2007	2008	2009	2010
1. Number of new conditions evaluated as red by the Reactor Oversight Process is ≤ 3.	0	0	0	0	0	0
2. Number of significant accident sequence precursors of a nuclear reactor accident is zero.	0	0	0	0	0	0
3. Number of operating reactors with integrated performance that entered the Inspection Manual Chapter 0350 process, or the multiple/repetitive degraded cornerstone column, or the unacceptable performance column of the Reactor Oversight Process Action Matrix, with no performance exceeding Abnormal Occurrence Criteria is ≤ 3.	0	0	1	0	0	0
4. Number of significant adverse trends in industry safety performance with no trend exceeding the Abnormal Occurrence Criterion I.D.4 is ≤ 1.	0	0	0	0	0	0
5. Number of events with radiation exposures to the public and occupational workers that exceed Abnormal Occurrence Criterion I.A.3:						
Reactors: 0.	0	0	0	0	0	0
Materials: ≤2.	1	0	0	0	0	0
Waste: 0.	0	0	0	0	0	0
6. Number of radiological releases to the environment that exceed applicable regulatory limits:						
Reactor: ≤0.	0	0	0	0	0	0
Materials: ≤2.	0	0	0	0	0	0
Waste: 0.	0	0	0	0	0	0

Analysis of FY 2010 Results

1. **Reactor Oversight Process:** The NRC reactor inspection program monitors nuclear power plant performance in three areas: (1) reactor safety, (2) radiation safety, and (3) security. Analysis of plant performance is based on many performance indicators and inspection findings. Each finding is then sorted into one of four categories: green, white, yellow, or red. Red indicates findings of high safety significance. There were no red performance indicators or findings in FY 2010.

2. **Reactor significant precursors:** This statistical measure of risk determines the likelihood of an event adversely impacting safety. A significant precursor is an event that has a probability of 1 or greater in 1,000 of leading to substantial damage to the reactor fuel. The NRC has identified no significant precursor events, based on screening reviews.

3. **Reactor performance:** The conditions in this measure indicate whether the NRC finds significant performance issues in a plant during an inspection or based on performance indicators under the Reactor Oversight Process. If any of the conditions in this measure are not met, the NRC will take action to ensure that plant safety is improved. No reactors met the conditions in this measure in FY 2010.

4. **Reactor safety trends:** This measure tracks trends for several key indicators of industry safety performance. These indicators provide insights into major areas of reactor performance, including reactor safety, radiation safety, and emergency preparedness. Statistical analysis techniques are applied to each indicator to calculate long-term trends. These trends represent industry averages rather than individual plant performance. No statistically significant adverse trends have been identified in any of the indicators in FY 2010.

5. **Nuclear material radiation exposures:** This measure tracks the number of radiation exposures to the public and occupational workers that exceed Abnormal Occurrence Criterion I.A.3, which is defined as those events that produce unintended permanent functional damage to an organ or a physiological system, as determined by a physician. This measure tracks both nuclear reactors and other nuclear material users, such as hospitals and industrial users. There were no events that met Abnormal Occurrence Criterion I.A.3 in FY 2010.

6. **Nuclear material releases to the environment:** This measure indicates the effectiveness of the NRC's nuclear material environmental regulatory programs. Exceeding the applicable regulatory limits is defined as a release of radioactive material that causes a total effective radiation dose equivalent to individual members of the public greater than 0.1 roentgen equivalent man (REM) in a year, exclusive of dose contributions from background radiation. No nuclear material releases to the environment that exceeded regulatory limits occurred in FY 2010.

Nuclear Safety Programs

The NRC engages in a comprehensive regulatory program that oversees the activities of its licensees. The core of its regulatory program is its licensing and oversight activities. The next sections describe the safety programs the NRC undertook during FY 2010 that enable it to achieve its Safety goal, strategic outcomes, and performance measure targets. The programs include: operating reactors, new reactors, fuel facilities, nuclear material users, high-level waste repository, spent fuel storage and transportation, decommissioning and low-level waste, as well as research activities, emergency preparedness and incident response, and international activities.

The Industry Trends Program

The NRC measures the effectiveness of its Nuclear Reactor Safety program activities based on the continued safe operation of the Nation's nuclear power plants. The NRC compiles data on overall safety performance using several industry-level performance indicators, a number of which are described in the following pages. These indicators show significant improvement in the long-term trends for safety performance of nuclear power plants. Plant operating experience data have yielded a steady stream of improvements in the reliability of plant systems and components, plant operating procedures, training of power plant operators, and regulatory oversight. For ease of viewing, all of the charts in this section display data since 1993.

The industry safety indicators are derived through engineering and scientific analyses by the NRC's Office of Nuclear Reactor Regulation and Office of Nuclear Regulatory Research. The performance indicator results are subject to minor variations as licensees submit revisions to the source data and may differ slightly from data reported in previous years as a result of refinements in data quality. Since the final data is not available until February of each year, this report will only show final fiscal year data from FY 1993 - 2009. The results of these analyses are reported annually to both the Commission and to Congress.

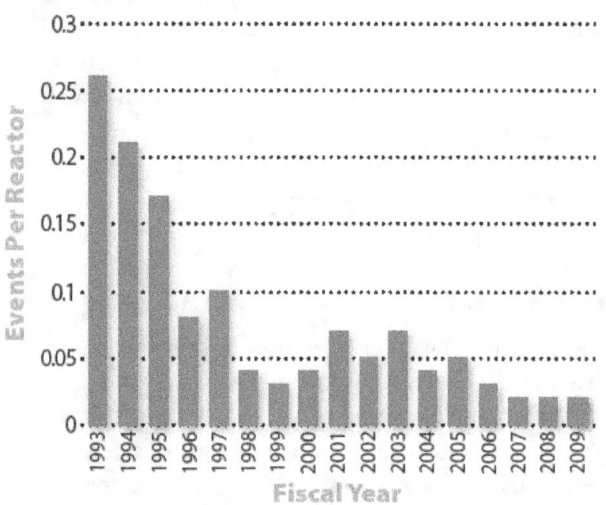
Figure 10
SIGNIFICANT EVENTS

Significant events meet specific criteria such as degradation of important safety equipment. The agency reviews operating events and assesses their safety significance. The number of significant events has declined since 1993.

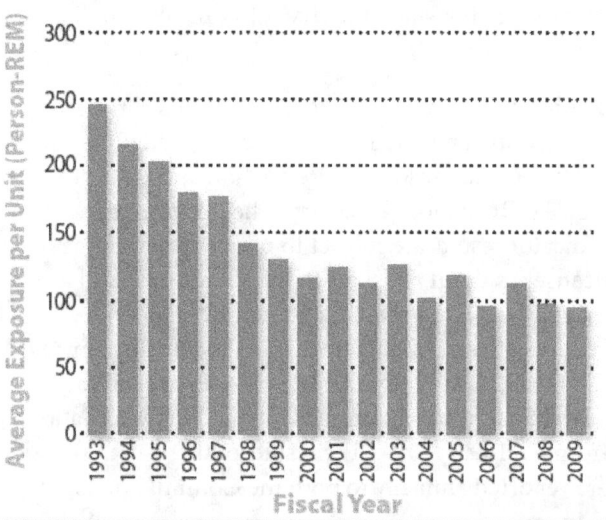
Figure 11
RADIATION EXPOSURE

The total (collective) radiation dose received by workers is an indication of the radiological challenges of maintaining and operating nuclear power plants. The trend shows a reduction in collective dose and demonstrates the effectiveness of the controls on radiation exposure implemented to meet these challenges

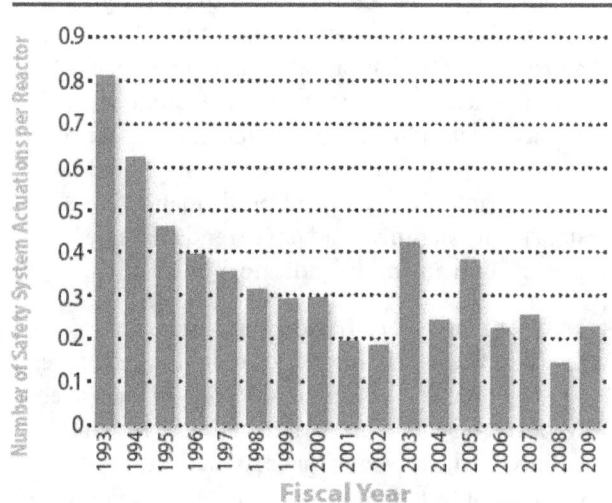
Figure 12
SAFETY SYSTEM ACTUATIONS

Safety systems mitigate off-normal events such as the widespread power blackout in August 2003, by providing reactor core cooling and water addition. Actuations of safety systems that are monitored include certain emergency core cooling and emergency electrical power systems. Actuations can occur as a result of "false alarms" (such as testing errors) or in response to actual events.

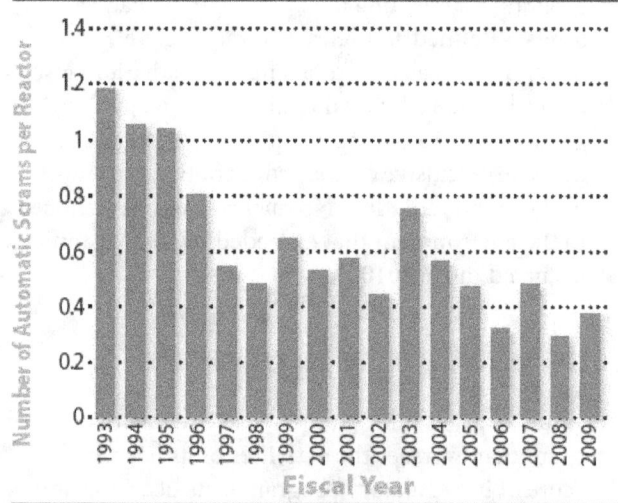
Figure 13
AUTOMATIC SCRAMS

A scram is a basic reactor protection safety function that shuts down the reactor by inserting control rods into the reactor core. Scrams can result from events that range from relatively minor incidents to precursors of accidents. The massive power blackout in August 2003 accounts for most of the increase in FY 2003, but has not affected the statistical trend for number of scrams, which has been declining steadily.

Figure 14
PRECURSOR OCCURRENCE RATE

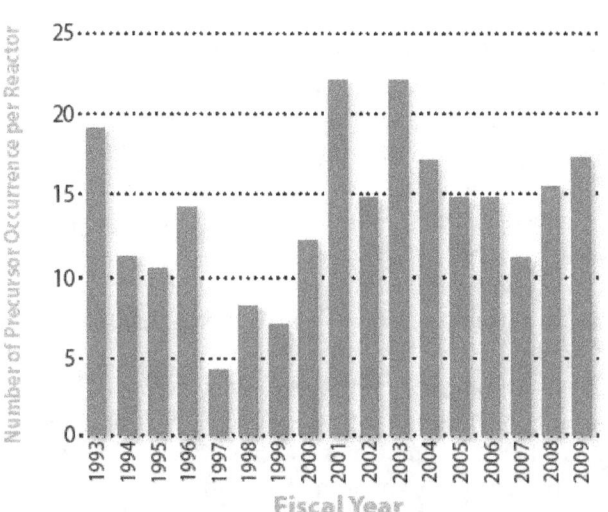

A precursor event is an event that has a probability of greater than 1 in 1 million of leading to substantial damage to the reactor fuel. There is no statistically significant adverse trend in the occurrence rate of precursor events since 1993, the baseline year for the statistical analysis. In addition, no statistically significant trend is detected for all precursors during the FY 2001–2009 period. Due to the complexities associated with evaluating precursor events, the data always lag behind other indicators.

Figure 15
SAFETY SYSTEM FAILURES

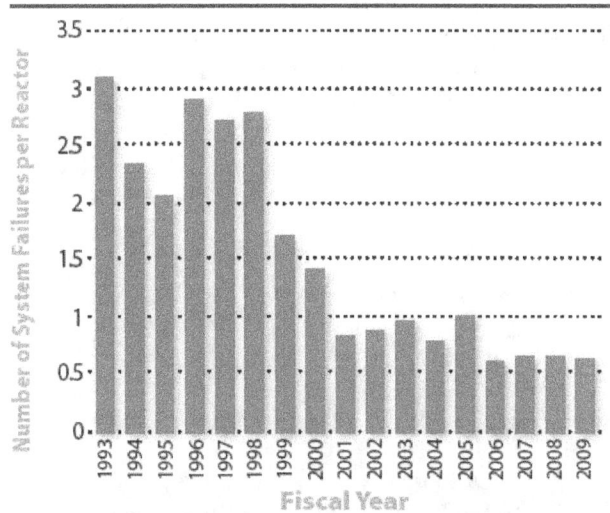

Safety system failures include any events or conditions that could prevent a safety system from fulfilling its safety function. The statistical trend for number of safety system failures across the industry has been declining.

Operating Reactors

Nuclear Reactor Licensing Activity

The agency's nuclear reactor licensing activity ensures that civilian nuclear power reactors and research and test reactors are operated in a manner that adequately protects public health and safety and the environment while safeguarding special nuclear material used in nuclear reactors.

The NRC completed 988 reactor licensing actions in FY 2010 (see Figure 16). The number of completed licensing actions has declined since 2007 because of a significant decrease in the number of licensing actions submitted to the agency. From FY 2003 through

FY 2007 the security enhancement requirements for licensees, in response to the terrorist attacks of September 11, 2001, resulted in an increase in licensing action submittals by licensees. The decrease in the number of licensing actions since 2007 is the result of the security enhancements being implemented by licensees. The NRC does not expect licensing action submittals to return to the elevated levels of FYs 2001–2007.

During FY 2010, the NRC completed 93 percent of the licensing actions in the agency's inventory within 1 year of receipt and 100 percent within 2 years (see Figure 18, page 31).

Watts Bar Unit 1 received a full power operating license in early 1996, and is presently the last power reactor to be licensed in the United States. The Tennessee Valley Authority (TVA) suspended construction of Watts Bar Unit 2 in 1985. In August 2007, TVA informed the NRC of its plan to resume construction of Watts Bar Unit 2. In FY 2010, the NRC continued its review of the operating license application, which TVA updated in March 2009. The NRC is proceeding with its reviews of safety, environmental, physical security, and emergency preparedness. In FY 2010, the NRC also assigned dedicated resident inspectors to monitor TVA's construction activities.

Power Uprates

The NRC also evaluates nuclear reactor power uprate applications, which allow licensees to safely increase the power output of their plants. The NRC review focuses on the potential impacts of the proposed power uprate on overall plant safety and confirms that plant operation at the increased power level is safe. During FY 2010, the NRC completed two power uprate licensing actions and met its established timeliness goals. The cumulative additional electric power from all power uprates approved since 1977 is approximately 5,726 megawatts. The NRC currently has 16 power uprates under review comprising a total of approximately 1,145 megawatts of electric power. Collectively, these uprates have added generating capacity at existing plants that is equivalent to more than 5 new reactors. The NRC expects to receive 39 new power uprate applications in the next 5 years totaling about 2,419 megawatts of electric power.

License Renewal

The NRC grants reactor operating licenses for 40 years, which can be renewed for an additional 20 years. The review process for renewal applications is designed to assess whether a reactor can continue to be operated safely during the extended period.

Figure 16
LICENSING ACTIONS COMPLETED

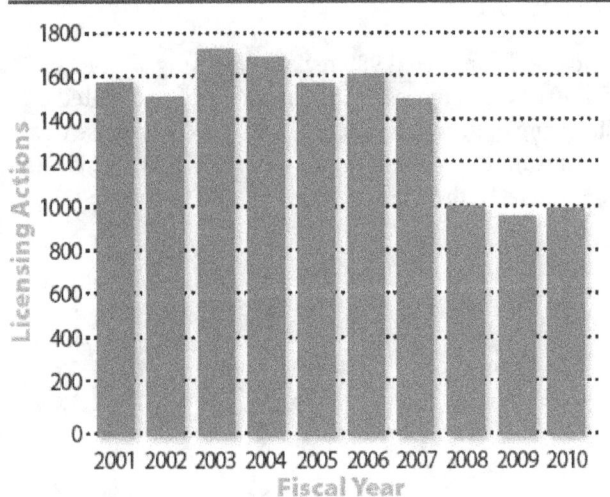

Figure 17
LICENSE RENEWAL APPLICATIONS

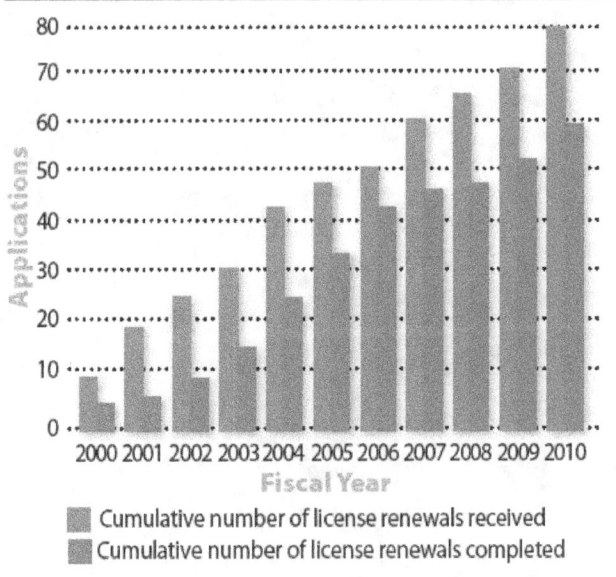

The NRC has received applications to renew the licenses for 80 units at 48 sites since the license renewal program began in 2000; it has renewed licenses for 59 units at 34 sites during that time (see Figure 17). The NRC is currently reviewing applications to renew the licenses for 21 units at 14 sites. The agency expects that almost all of the licensees for currently licensed units will eventually apply to renew their licenses.

Figure 18
LICENSING ACTION AGE

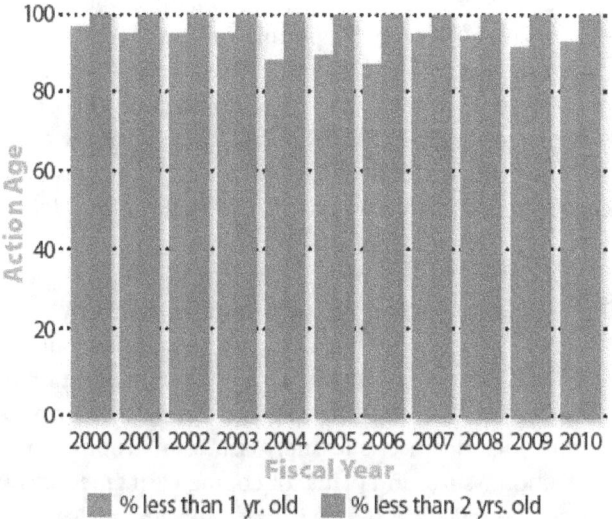

% less than 1 yr. old   **% less than 2 yrs. old**

Nuclear Reactor Inspection

The NRC provides continuous oversight of nuclear reactors through the Reactor Oversight Process (ROP) to verify that nuclear plants are operated safely and in accordance with the agency's rules and regulations. The NRC performs a rigorous program of inspections at each plant and may perform supplemental inspections and take additional actions to ensure that the plants address significant safety issues. The NRC has full authority to demand that a licensee take immediate action for any conditions that result in excess risk to the public, including requiring a plant to shut down if necessary. The NRC also conducts public meetings with licensees to discuss the results of the agency's assessments of its safety performance.

The agency evaluates both inspection findings and performance indicators to assess the performance of each operating nuclear power plant. In FY 2010, all of the Nation's nuclear power plants operated safely. The safety indicators for nuclear plants as a whole showed no adverse trends, and more than 99 percent of plant safety indicators were rated green, which is the highest safety rating. The results of NRC inspection findings for each plant are documented in inspection reports and are available to the public at http://www.nrc.gov/NRR/OVERSIGHT/ASSESS/pim_summary.html.

The NRC assesses its inspection program on a regular basis. Assessments conducted in FY 2010 confirm that the agency's ROP met its goal of conducting an objective, risk-informed, and predictable regulatory process that focuses NRC and licensee resources on aspects of plant performance that have the greatest impact on safe plant operations. More information on reactor inspection is available at http://www.nrc.gov/reactors/operating.html.

Rulemaking

During FY 2010, the NRC undertook several important rulemaking activities to improve protection of public health and safety and the environment and enhance the effectiveness of its regulations. The NRC published a final rule to provide alternate requirements for protection against pressurized thermal shock (PTS) events in reactor vessels, using updated analysis methods. This rule allows licensees of operating pressurized water reactors to adopt a more realistic approach for determining the probability of vessel failure during a PTS event. Further, the agency published a proposed rule to obtain fingerprint-based background checks for staff with unescorted access to research and test reactors. Finally, the agency published a proposed rule updating NRC requirements for the generic environmental impact statement (GEIS) that addresses the environmental effects of renewing power reactor operating licenses. This proposed rule redefines the number and scope of the environmental issues that must be addressed as part of a license renewal application and those that may be addressed generically.

Reactor Investigations and Enforcement

Compliance with NRC requirements plays an important role in ensuring that safety is being maintained. NRC policies deter noncompliance and encourage prompt identification and timely, comprehensive corrective actions. Licensees, contractors, and their employees who do not achieve the high standard of compliance expected by the NRC are subject to enforcement sanctions and investigations of potential willful violations. Each enforcement action depends on the circumstances of the case. The NRC will not permit licensees to continue to conduct licensed activities if they cannot achieve and maintain adequate levels of safety. In FY 2010, the NRC issued 39 escalated enforcement actions, one of which involved a civil penalty totaling $70,000 in proposed fines. Escalated enforcement actions include all notices of violation (NOV) categorized at severity level of I, II, or III; those NOVs associated with a white, yellow or red finding as categorized by the Reactor Oversight Process; and all enforcement related orders.

New Reactors

The NRC reviews applications for new reactor facilities submitted by prospective licensees and issues standard design certifications, early site permits, limited work authorizations, construction permits, operating licenses, and combined operating licenses (COL) when appropriate. At present, the NRC anticipates that these activities will involve new light-water reactor (LWR) facilities in a variety of projected locations throughout the United States.

Design Certification

The NRC is reviewing three design certifications and two design certification amendments. By issuing a design certification, the NRC approves a nuclear power plant design independent of an application to construct or operate a plant. A design certification is valid for 15 years from the date of issuance, but can be renewed for an additional 10 to 15 years.

The NRC is currently reviewing design certifications for General Electric's Economic Simplified Boiling-Water Reactor, AREVA's Evolutionary Power

Reactor, and Mitsubishi's U.S. Advanced Pressurized-Water Reactor. The agency is also in the process of reviewing design certification amendments for the Westinghouse AP1000 and the South Texas Advanced Boiling Water Reactors to address the requirements in the Commission's new rule Title 10 of the *Code of Federal Regulations* (10 CFR) 50.150, "Aircraft Impact Assessment."

The NRC conducted a Lean Six Sigma review of the design certification rulemaking process during FY 2010 to improve the effectiveness and efficiency of the process. The agency further enhanced the rulemaking process by beginning rulemaking activities earlier than previously planned. These improvements reduced impediments to making timely decisions on new reactor license applications that reference the designs being certified, while still including review and comments from all internal and external stakeholders.

Early Site Permits

The NRC approves the site for a nuclear facility by issuing an early site permit. Early site permits are valid for 10 to 20 years and can be renewed for an additional 10 to 20 years. The NRC review of an early site permit application addresses site safety issues, environmental protection issues, and plans for coping with emergencies, independent of the review of a specific nuclear plant design. The agency issued early site permits to the Clinton site in Illinois in March 2007, the Grand Gulf site in Mississippi in April 2007, the North Anna site in Virginia in November 2007, and the Vogtle site in Georgia in August 2009.

In March 2010, the NRC received an early site permit application from Exelon Nuclear Texas Holdings for the Victoria County Station site located in Victoria County, TX. The agency finished its acceptance review of the Victoria County Station early site permit in June 2010. In May 2010, the NRC received an early site permit application from the Public Service Enterprise Group for a site adjacent to the Salem and Hope Creek Generating Stations now operating in Lower Alloways Creek, Salem County, NJ. The agency completed its acceptance review of the Public Service Enterprise Group early site permit in August 2010.

Combined Operating License

A combined operating license (COL) authorizes construction and operation of a nuclear power plant. The application for a COL must contain essentially the same information required in an application for an operating license, including financial and antitrust information and an assessment of the need for power. The application must also describe the inspections, tests, analyses, and acceptance criteria (ITAAC) that are necessary to ensure that the plant has been properly constructed and will operate safely.

The NRC has two objectives for the review of COL applications. The first objective is to ensure that the proposed new reactor designs and planned operations will be in accordance with NRC regulations for safety, security, and protection of the environment. The second objective is that the reviews will be completed on the schedules negotiated with applicants.

For FY 2010, the agency established a target to complete milestones associated with conducting up to 20 COL application reviews. Since 2007, the agency has docketed all 18 COL applications received for sites across the country. The agency is actively reviewing 13 of the 18 applications. Applicants have withdrawn or asked the agency to suspend reviews of five applications: Grand Gulf, Victoria County, Callaway, Nine Mile Point, and River Bend. Victoria County withdrew its COL application and submitted an early site permit application in FY 2010. The agency did not receive any new COL applications in FY 2010.

The NRC issued the final supplemental environmental impact statements for the North Anna COL application and the draft environmental impact statements for the following applications: South Texas, V.C. Summer, Calvert Cliffs, Levy, and Comanche Peak. Issuance of the draft environmental impact statement is a major milestone in the environmental review for COLs because this document is issued for public comment and reviewed by the U.S. Environmental Protection Agency.

The NRC has developed a construction inspection program for plants to be licensed under 10 CFR Part 52, "Licenses, Certifications, and Approvals for Nuclear Power Plants," and undertook many critical program development activities in FY 2010. For example, the agency produced a number of draft and final construction inspection program materials, such as inspection procedures, inspection strategy documents, regulatory guides, Inspection Manual chapters, and a construction inspection program information brochure for stakeholders in both English and Spanish. The staff developed a draft approach for maintenance of completed ITAAC and continued developing a detailed ITAAC closure verification process. The NRC staff also continued development of: (1) inspector training, (2) business processes to support additional identified information technology (IT) system needs, (3) generic inspection schedules, and (4) enhancements to the existing assessment and enforcement program for new reactors. In addition, the NRC maintained an aggressive schedule of public meetings to provide a forum for stakeholders to participate and comment on staff proposals for ITAAC closure, licensee assessment, enforcement, and other construction inspection program topics.

The NRC maintains a regular schedule of vendor inspections and an active program of international cooperation to support increased fabrication activities domestically and internationally in response to new reactor construction plans. The agency conducts these inspections to ensure the effective implementation of quality assurance program requirements imposed on vendors by NRC applicants and licensees. The agency conducts a minimum of 10 domestic and international vendor inspections per year. In FY 2010, the NRC completed 13 inspections.

Related international cooperative efforts have included multinational vendor inspections, technical discussions with foreign regulatory counterparts, sharing vendor experience and other information with other countries, NRC inspector rotations to facilities under construction in other countries, and participation in the Vendor Inspection Cooperation Working Group under the auspices of the Multinational Design Evaluation Program (MDEP). Exchanges such as these have provided key insights into each country's methods of oversight and have enabled the agency to build a foundation of trust and a rapport for communicating and sharing key information and findings.

In FY 2010, the NRC continued to enhance the regulatory framework for COLs to clarify requirements for licensees. The NRC issued the following six interim staff guidance (ISG) documents for COLs: (1) ISG-10 "Review of Evaluation To Address Adverse Flow Effects in Equipment Other Than Reactor Internals," (2) ISG-11 "Finalizing Licensing Basis Information," (3) ISG-15 "Post-Combined License Commitments," (4) ISG-16 "Staff Guidance on Interim Guidance DC/COL-15G-016 with 10 CFR 50.54 (hh) (2) and 10 CFR 52.80 (d)," (5) ISG-17 "Ensuring Hazard-Consistent Seismic Input for Site Response and Soil Structure Interaction Analyses," and (6) ISG-20 "Seismic Margin Analysis for New Reactors Based on Probabilistic Risk Assessment."

Advanced Reactor Program

The NRC has continued its efforts to support congressionally mandated and U.S. Department of Energy (DOE)-sponsored programs such as the Next Generation Nuclear Plant, while also supporting efforts related to the growing commercial interest in integral pressurized-water reactors. The agency has also focused—and continues to focus—on the identification and resolution of generic policy issues as well as key technical issues for the licensing of small modular reactor (SMR) designs while concurrently training its staff to be prepared for the review of potential future SMR applications.

During FY 2010, the NRC continued its strong outreach to conduct preapplication interactions with stakeholders and four potential applicants. The agency hosted workshops focusing on potential policy and technical issues and the 10 CFR Part 52 licensing process and plans to host a workshop on manufacturing licenses later in 2010. In addition, the

agency staff issued a regulatory issue summary asking for voluntary responses from companies interested in submitting applications for SMRs to help effectively plan resources.

Fuel Facilities

Licensing

The NRC licenses and inspects all commercial nuclear fuel facilities that process and fabricate uranium concentrates into the reactor fuel that powers the Nation's nuclear reactors. Licensing activities include detailed health, safety, safeguards, and environmental licensing reviews of licensee programs, procedures, operations, and facilities to ensure safe and secure operations.

The NRC completed significant fuel cycle licensing reviews during FY 2010. Throughout the year, the agency completed a high volume of license amendments and other licensing reviews to support initial operations of the URENCO USA (formerly LES) enrichment facility in Eunice, NM. After conducting a thorough operational readiness review between December 2009 and June 2010, the agency approved operation of the first centrifuge cascade in June 2010. This is the first new uranium enrichment facility in the United States since 1954. When operating at full capacity, the facility could supply the enrichment needs of about one-fourth of the operating commercial nuclear power plants in the country.

The NRC also completed development of the draft safety evaluation report on the license application for the Mixed Oxide Fuel Fabrication Facility under construction at the Savannah River site near Aiken, SC. This facility is designed to process 34 metric tons of plutonium from the nuclear weapons stockpile into mixed-oxide (plutonium and uranium) fuel for use in commercial nuclear power plants. In accordance with the agency's regulations, the NRC will not issue its decision about whether to license the Mixed Oxide Fuel Fabrication Facility until the agency verifies completion of the principal structures, systems, and components of the facility. The facility is estimated to be completed in 2016.

The NRC continued the safety, security, and environmental reviews of two license applications for uranium enrichment facilities. Uranium enrichment facilities increase the concentration of the uranium-235 isotope from its natural enrichment of about 0.7 percent of natural uranium to 4 to 5 percent for use in commercial LWRs, such as those used throughout the commercial power industry in the United States. AREVA submitted an application in December 2008 to build a centrifuge enrichment facility near Idaho Falls, ID. Another application, submitted in June 2009 by General Electric-Hitachi, is for a laser-based enrichment facility to be built in Wilmington, NC. The agency completed draft environmental impact statements and conducted public outreach on its environmental review for both enrichment facilities during FY 2010.

A byproduct of uranium enrichment is depleted (i.e., reduced in the uranium-235 isotope) uranium hexafluoride. During FY 2010, the agency also accepted a license application to construct and operate a facility to convert depleted uranium hexafluoride into an oxide form for ultimate disposal and to recover the fluorine from the uranium hexafluoride for other commercial applications.

Oversight

The NRC's fuel cycle oversight process consists of both planned and reactive inspections with enforcement and periodic assessments based on the findings of these inspections. The NRC has full authority to demand that a licensee take immediate action for any conditions that result in excess risk to the public, including requiring the facility to shut down.

The NRC conducted a thorough review of the root and contributing causes of an event that occurred in October 2009 at the Nuclear Fuel Services facility in Erwin, TN. Although the event did not cause a release of hazardous material to the environment and had no public health and safety consequences, the agency's review of the underlying causes and the licensee's response to the event led to a shutdown of all licensed processes at the facility in December

2009 and NRC issuance of a confirmatory action letter in January 2010. This confirmatory action letter established the corrective measures to be taken by the licensee before seeking agency approval to restart process lines. As the licensee identified its readiness to restart each process line, the agency conducted additional inspections to verify readiness for restart and supplemental inspections during the initial operation of each process line as it was restored to service.

Rulemaking

In response to sustained industry interest in potential reprocessing of spent nuclear fuel, the NRC continued to work on developing a technical basis for rulemaking to establish the regulatory framework for licensing a reprocessing facility. In 2009, the agency completed a review to identify and prioritize gaps in the existing regulations. During FY 2010, the agency continued to define the technical basis needed to support the development of proposed regulations to resolve the identified gaps and establish an effective and efficient regulatory framework.

Fuel Facility Investigation and Enforcement

The NRC will not permit licensees to continue to conduct licensed activities if they cannot achieve and maintain adequate levels of safety and security. The agency assesses compliance, takes enforcement, and investigates potential willful violations. For fuel facilities in FY 2010, the NRC issued 14 escalated enforcement actions, some of which involved civil penalties. Of these, six involved civil penalties totaling $223,750 in proposed fines. Escalated enforcement actions include all notices of violation (NOV) categorized at a severity level of I, II, or III; and all enforcement-related orders.

Allegations of fuel facility-related wrongdoing are referred to the NRC Office of Investigations for appropriate action. The Office of Investigations (OI) actively investigates allegations of fuel facility-related wrongdoing. In FY2010, OI conducted 16 fuel facility-related investigations. Allegations included, but

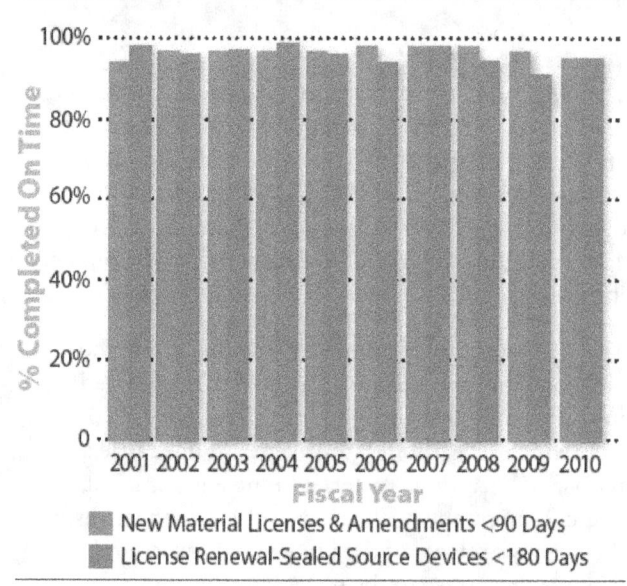

U.S.NRC
UNITED STATES NUCLEAR REGULATORY COMMISSION

were not limited to: improper handling of classified information, providing incomplete and inaccurate information to the NRC, tampering with weapons, falsification of inspection reports, and discrimination for raising safety concerns. The agency has referred all substantiated investigations to the U.S. Department of Justice for prosecution.

Nuclear Material Users

The NRC licenses and inspects the commercial use of nuclear material for industrial, medical, and academic purposes. Commercial uses of nuclear material include medical diagnosis and therapy, medical and biological research, academic training and research, industrial gauging and nondestructive testing, production of radiopharmaceuticals, and fabrication of commercial products (such as smoke detectors) and other radioactive sealed sources and devices. The agency currently regulates more than 2,980 specific licensees for the use of nuclear byproduct and other radioactive materials. Under the NRC's Agreement State program, 37 States have assumed primary regulatory responsibility over the industrial, medical, and other users of nuclear materials in their States. The NRC works closely with these States to ensure that they maintain public safety.

Detailed health and safety reviews of license applications, as well as inspections of licensee procedures, operations, and facilities, provide reasonable assurance of safe operations and the production of safe products. The NRC routinely inspects nuclear material licensees to ensure that they are using nuclear materials safely, maintaining accountability of those materials, and protecting public health and safety. The agency also analyzes operational experience from NRC and Agreement State licensees and regularly evaluates the safety significance of events reported by licensees and Agreement States.

These States have entered into agreements under Section 274 of the Atomic Energy Act, as amended, that provides for the State to assume regulatory authority for sources, byproducts, and limited

quantities of special nuclear material. These Agreement States regulate approximately 19,600 licensees (academic, medical and industrial uses). No States are actively pursuing a new agreement at this time. The NRC reviews the Agreement State programs as well as certain NRC licensing and inspection programs through the Integrated Materials Performance Evaluation Program.

Licensing and Inspection

The NRC completed 2,460 materials licensing actions and 930 routine health and safety inspections in FY 2010. The agency maintained its high standards with timely reviews of nuclear material license renewals and sealed-source and device designs in FY 2010. The agency completed 95 percent of new application and license amendment reviews within 90 days of receipt and 95 percent of license renewal and sealed-source and device design reviews within 180 days of receipt (see Figure 19).

Figure 19

TIMELINESS REVIEW OF NUCLEAR MATERIAL LICENSING APPLICATIONS

Rulemaking

In FY 2010, the NRC undertook several rulemaking activities to allow the use of radioactive materials while protecting public health and safety and the environment. These activities included publication of a proposed rule to enhance domestic nonproliferation activities in accordance with International Atomic Energy Agency (IAEA) recommendations. The NRC is also proposing to amend its regulations that govern the licensing and distribution of byproduct materials to make the regulations clearer, more risk-informed, and up-to-date. Additionally the NRC is proposing to modify the definition of medical events for permanent implant brachytherapy, which will facilitate the ability of medical licensees to recognize medical events earlier. The agency also published a final rule related to the requirements for categorical exclusion from environmental review under the National Environmental Policy Act of 1969.

Investigation and Enforcement

The NRC will not permit licensees to continue to conduct licensed activities if they cannot achieve and maintain adequate levels of safety. The agency assesses compliance, takes enforcement, and investigates potential willful violations. For nuclear material users in FY 2010, the NRC issued 71 escalated enforcement actions, some of which involved civil penalties. Of these, four involved civil penalties totaling $284,700 in proposed fines. Escalated enforcement actions include all notices of violation (NOV) categorized at a severity level of I, II, or III; and all enforcement-related orders. These actions were assessed to material user licensees, their contractors, or individuals.

In FY 2010 the Office of Investigations (OI) conducted 40 nuclear materials-related investigations. Allegations included, but were not limited to: failure to comply with storage requirements, discrimination for engaging in protected activity, falsification of inspection reports, and providing incomplete and inaccurate information to the NRC. The agency has referred all substantiated investigations to the U.S. Department of Justice for prosecution.

High-Level Waste Repository

The high level waste repository program encompasses the NRC's licensing activities related to DOE's proposed Yucca Mountain geologic repository. This program supports the agency's responsibilities associated with the licensing review of the DOE application for the permanent disposal of spent fuel at Yucca Mountain, NV. To conduct the license application review, the agency implemented two concurrent processes. The first process is to assess the technical merits of the repository design. The second process is to support the adjudicatory hearing before the NRC Atomic Safety and Licensing Boards convened to hear the technical and legal challenges in the adjudicatory proceeding.

On September 8, 2008, the NRC docketed the June 3, 2008 application from DOE, for a license to construct and operate the nation's first geologic repository for high-level nuclear waste at Yucca Mountain, NV. This initiated the NRC staff's review of the technical merits of the repository application and formulation of a position on whether to issue a construction authorization for the repository. In May 2009, the Atomic Safety and Licensing Board granted petitions to intervene regarding the DOE license application and admitted contentions. On March 3, 2010, DOE filed a motion to withdraw its license application, with prejudice. On June 29, the Licensing Board denied DOE's motion. On June 30, the Commission invited briefing by the parties as to "whether the Commission should review, and reverse or uphold, the Board's decision." The briefing was completed on July 16, 2010, and the case is pending before the Commission.

In FY 2010, the staff has continued to conduct a technical review of the application and issued the first of five volumes of NUREG-1949, "Safety Evaluation Report Related to Disposal of High-Level Radioactive Wastes in a Geologic Repository" at Yucca Mountain, NV, which documents the results of the NRC staff's review of the general information that DOE provided in its 2008 repository application.

Spent Fuel Storage and Transportation

The NRC ensures that spent fuel is safely stored and transported. The agency conducts licensing and certification reviews to ensure that interim spent fuel storage facility and cask designs and domestic and international shipments of spent fuel and other risk-significant radioactive materials are safe and secure and comply with agency regulations.

Shipments of radioactive materials are safely and securely transported each year within the United States. Several Federal agencies share responsibility for regulating the safety and security of those shipments. The NRC closely coordinates its transportation-related activities with those of the U.S. Department of Transportation and, as appropriate, DOE. The agency inspects vendors, fabricators, and licensees that use transport packages, spent fuel storage casks, and interim storage of spent fuel both at and away from reactor sites to help ensure the safety and security of spent fuel storage and transportation.

Licensing and Inspection

In FY 2010, the NRC completed 59 transport package design reviews and 19 storage cask and facility design reviews. The review of transportation and interim storage licensing requests ensures that shipments are made in NRC-approved packages that meet rigorous performance requirements and verifies that spent fuel is safely stored, thereby enabling continued reactor and decommissioning operations. The agency also conducted 20 inspections of activities related to radioactive material package certificate holders, spent fuel storage cask certificate holders, and inspections at independent spent fuel storage facilities to ensure that casks are being designed, fabricated, and used according to approved safety requirements.

Rulemaking

In FY 2010, the NRC completed rulemaking changes to its regulations concerning licensing requirements for the independent storage of spent nuclear fuel, high-level radioactive waste, and

reactor-related greater than Class C waste. The amendments extend and clarify the license terms for dry storage cask certificates of compliance (CoCs) and independent spent fuel storage installation licenses. The amendments also require certain aging management requirements for both specific license and CoC renewals. Finally, the amendments allow general licensees under 10 CFR Part 72, "Licensing Requirements for the Independent Storage of Spent Nuclear Fuel, High-Level Radioactive Waste, and Reactor-Related Greater than Class C Waste," to implement changes authorized by a later CoC amendment to a cask loaded under the initial CoC or an earlier CoC amendment. This rulemaking is needed to improve the regulatory efficiency of 10 CFR Part 72. The final rule will be issued and become effective in FY 2011.

The NRC developed a plan for integrating spent nuclear fuel regulatory activities to more effectively address the regulatory and licensing aspects of extended storage and transportation, reprocessing, and disposal of spent nuclear fuel and high-level waste. The purpose of the plan is to ensure that regulation of the back end of the fuel cycle accomplishes safety, security, and environmental protection in an efficient and effective manner and that decisions made about one component or area of this system adequately consider other components or areas (i.e., treating spent fuel and high-level waste regulation as a system of interrelated activities). By integrating the approach for regulation of spent nuclear fuel or high-level waste, the agency can improve the efficiency and effectiveness of its regulatory processes and gives stakeholders stability and predictability in a dynamic environment.

The NRC also began a comprehensive review of the spent fuel storage and transportation regulatory programs to evaluate their adequacy for ensuring safe and secure storage of spent fuel for extended periods beyond 120 years, including research to bolster the technical bases of the regulatory framework in support of extended periods.

Decommissioning and Low-Level Waste

Decommissioning removes radioactive contamination from buildings, equipment, ground water, and soil, achieving levels that permit the release of the property while protecting the public. The NRC terminates the licenses for decommissioned facilities after the licensees demonstrate that the residual onsite radioactivity is within regulatory limits and sufficiently low to protect the health and safety of the public and the environment. Completion of decommissioning, environmental, and performance assessment activities enables sites to return to productive use while ensuring that residual radioactivity does not pose an unacceptable risk to the public. Agreement States ensure that the licensees in their jurisdiction properly decommission their facilities in accordance with State regulations, which must be compatible with NRC regulations.

Decommissioning

In FY 2010, the NRC oversaw decommissioning activities at approximately 85 power and early demonstration reactors, research and test reactors, uranium recovery sites, complex materials sites, and fuel cycle facilities. The agency increased its emphasis on the decommissioning of legacy uranium recovery sites during FY 2010 and has worked extensively with the U.S. Environmental Protection Agency, the State of New Mexico, and Native American Tribes on decommissioning activities at the United Nuclear Corporation Churchrock, Homestake, and Ambrosia Lake Mill sites.

Uranium Recovery Licensing and Oversight

The NRC conducts regulatory oversight at eight operational uranium recovery sites and reviews and approves the applications for new, restarting, or expanding uranium recovery facilities. The agency reviewed 10 applications for new, restarting, expanding or decommissioning uranium recovery facilities received between FY 2007 and FY 2010. These reviews include initiating environmental reviews. The agency published a supplemental environmental impact statement in FY 2010 for one of those facilities, with two others projected to be completed in the first quarter of FY 2011. The agency also completed two

separate review activities related to the West Valley Demonstration Project Phase I Decommissioning Plan and the environmental decision. One was the technical evaluation report to show compliance with the environmental regulatory criteria. The other was the evaluation of compliance with the Commission's West Valley Demonstration Project Policy Statement. Both contributed to significant progress in decommissioning the site. Completion of these reviews, along with actions completed by the State of New York, allows the DOE to move forward with the long-stalled effort to decommission major portions of the West Valley site, including removal of the main processing plant structure and the source of the strontium-90 ground water plume.

Low-Level Waste

The NRC conducted regulatory activities to help ensure the safe management and disposal of low-level radioactive waste generated by radioactive material users, nuclear power plants, and other NRC licensees. The agency performed monitoring visits and issued reports for the Savannah River Site Saltstone facility and the Idaho National Laboratory. In addition, the agency has conducted outreach with stakeholders and licensees on issues related to the effect of the lack of low-level waste disposal options as a result of limited access to the Barnwell disposal facility.

Research Activities

The NRC's safety research program evaluates and resolves safety issues for nuclear power plants and other facilities and materials that the agency regulates. The agency conducts its research program to evaluate existing and potential safety issues; supply independent expertise, information, and technical judgments to support timely and realistic regulatory decisions; reduce uncertainties in risk assessments; and develop technical regulations and standards. When possible, the agency engages in cooperative research with other government agencies, the nuclear industry, universities, and international partners.

During the past year, the NRC research program addressed key areas that support the agency's safety mission. Some of the more important issues addressed include verification and validation of fire safety models; material degradation of reactor system and pressure boundary components, especially as it relates to license renewal periods; evaluation of digital systems for cyber vulnerabilities; seismic hazard issues; advanced reactor research; development of advanced tools for probabilistic risk assessment activities that support risk-informed regulatory decisionmaking; and severe reactor accident consequence analyses.

Fire Safety

The NRC continued to conduct collaborative research to develop state-of-the-art knowledge, guidance, methods, and tools in support of regulatory activities related to fire protection and fire risk analyses. This collaborative research included participation from the Electric Power Research Institute, the National Institute of Standards and Technology, Sandia and Brookhaven National Laboratories, and the University of Maryland. The NRC and the Electric Power Research Institute continue to provide training on NUREG/CR-6850, "EPRI/NRC RES Fire PRA Methodology for Nuclear Power Facilities," issued September 2005, for those nuclear power plants that have submitted letters of intent to transition to National Fire Protection Association Standard 805, "Performance-Based Standard for Fire Protection for Light Water Reactor Electric Generating Plants," via 10 CFR 50.48(c).

Advanced Reactor Research

In response to the Energy Policy Act of 2005, the NRC initiated research in a number of major technical areas related to licensing a prototype high-temperature gas-cooled reactor (HTGR), which can be used to generate electricity, hydrogen, or both. The agency developed HTGR preliminary plant models for incorporation into the NRC's safety analysis code, scoping analysis of important HTGR thermal-fluids phenomena using computational fluid dynamics tools, modification of light-water reactor specific reactor physics codes for HTGR nuclear analysis applications, and preliminary fuel performance models. The NRC also convened a meeting of international nuclear graphite experts to assess the knowledge gaps and participate in standards development activities. The agency has also begun to

generate models for its thermal-hydraulic and severe accident codes to support review of the new integral pressurized-water reactor (iPWR) small modular reactor designs. The agency developed generic iPWR models that can be used to explore postulated event sequences to support preapplication activities.

Materials Degradation

The NRC continues to research materials degradation issues for currently licensed reactors. The purpose of this research is to identify susceptible materials and assess component-specific degradation mechanisms in existing reactors to ensure continued safe operation. The agency is also performing research on reactor internals to determine the effects of neutron fluence and thermal effects on the physical properties of reactor internal materials. In addition, the agency is conducting research into potential technical issues that may challenge long-term safe operation of existing commercial nuclear power plants in second and subsequent license renewal periods.

Digital Instrumentation and Controls

The NRC's research supports the licensing of new digital instrumentation and control systems intended for use in retrofits to operating reactors and for use in new and next-generation reactors. The agency is also actively engaged in ongoing research on the evaluation of digital systems for cyber vulnerabilities. In FY 2010, the agency published Regulatory Guide 5.71, "Cyber Security Programs for Nuclear Facilities." This regulatory guide provides an approach that the NRC staff deems acceptable for complying with NRC regulations on the protection of digital computers, communications systems, and networks from a cyber attack as defined by 10 CFR 73.1, "Purpose and Scope."

Probabilistic Risk Assessment

The NRC continues to research the development of advanced models, methods, and tools for probabilistic risk assessment activities that support risk-informed regulatory decisionmaking. In FY 2010, the agency released Version 8 of the Systems Analysis Program for Hands-on Integrated Reliability Evaluations

(SAPHIRE) software that allows analysts to perform probabilistic risk assessments for any complex system, facility, or process. SAPHIRE supports the agency's risk-informed programs such as the Accident Sequence Precursor Program, the NRC's Incident Investigation Program, and the significance determination process. It is also used to develop and run the standardized plant analysis risk models.

Seismic Research

The NRC is researching seismic hazard issues to support the siting of new reactors and the evaluation of the seismic safety of existing nuclear facilities. In cooperation with academic institutions, other Federal and State agencies, and industry, the NRC is conducting a program to develop ground motion propagation and earthquake source zone models. The NRC is also conducting a study of potential tsunami sources and the resulting potential hazards to NRC-regulated facilities in collaboration with the U.S. Geological Survey and the National Oceanographic and Atmospheric Administration. The agency is using the results of this research to inform licensing decisions and update risk assessments.

State-of-the-Art Reactor Consequence Analysis

The State-of-the-Art Reactor Consequence Analysis (SOARCA) project involves the reanalysis of severe accident consequences to develop a body of knowledge about the realistic outcomes of severe reactor accidents. In addition to incorporating the results of over 25 years of research, the objective of the SOARCA study is to include in these updated plant analyses the significant plant improvements and updates (e.g., system improvements, training and emergency procedures, and offsite emergency response) that have been made by licensees. In FY 2010, the NRC completed a detailed technical evaluation of two types of commercial nuclear power plants. The draft report has been reviewed by an independent peer review panel of subject-matter experts and will be released for public review and comment before being finalized.

Emergency Preparedness and Incident Response

The NRC's emergency preparedness and incident response activities ensure that adequate measures can and will be taken to mitigate plant events and to minimize possible radiation doses to members of the public, and that the agency can respond effectively to events at licensee sites.

The NRC conducted many emergency exercises with its licensees and Federal partners in FY 2010. NRC emergency responders participated in 20 exercises with licensee sites across the country, four of which involved the NRC Headquarters response team. These exercises focused on licensee, State, and local responder implementation of onsite and offsite radiological emergency plans. The agency also used exercises to train its response organization and practice coordination activities with Federal partners, including the U.S. Department of Homeland Security (DHS). The agency participated in one hostile-action-based emergency preparedness drill, conducted voluntarily at the River Bend Station, and coordinated with the Federal Emergency Management Agency (FEMA) to observe many other hostile-action-based drills to better understand the unique challenges posed by hostile action events and to identify significant good practices and lessons learned.

In addition to exercises involving its licensees, the NRC participated in the annual continuity exercise (Eagle Horizon 10) for Federal Executive Branch departments and agencies, which included real-time relocation of the NRC's Continuity of Operations Plan management team and extended play. During April 27–29, 2010, four NRC staff members participated in the U.S. Environmental Protection Agency's Lead Liberty RadEx Cs 137 Radiological Dispersion Device Exercise in Philadelphia, PA. The NRC and FEMA also hosted a multiagency senior official tabletop exercise that focused on the challenges of aligning critical information and event communications related to reactor-accident and hostile-action-incident scenarios at a nuclear power plant.

The NRC is currently conducting a rulemaking that proposes to enhance the emergency preparedness regulations. Enhancements to the regulations include codifying voluntary industry efforts since September 11, 2001. The proposed rule was issued in the Federal Register on May 18, 2009. The NRC and FEMA also formed a joint comment resolution team to address cross-cutting issues, where comments pertain to both onsite and offsite emergency preparedness.

Consistent with its policy to provide States with potassium iodide as requested, the NRC worked with States to replenish potassium iodide supplies to be used as a supplement to public protective actions within the 10-mile emergency planning zones around nuclear power plants.

In FY 2010, the NRC continued to deploy a Web-based incident tracking system to improve functionality, enhance cyber security, and reduce operating costs. This system provides needed capabilities for response to multiple emergencies and enables rapid, accurate information sharing with NRC responders in the regions, at sites, or at home. The agency also continued its modernization of the Emergency Response Data System, which provides real-time information from nuclear power plants to the NRC and State operations centers during emergencies. The modernization of this system enhances cyber security and reliability and includes improvements to the user interface.

In FY 2010, the NRC finalized its pandemic plan and established a process for annual reviews required by the Homeland Security Council. Experiences during the influenza outbreaks in the spring and fall of 2009 helped to guide the planning and identification of needed actions, as well as increase confidence that the agency will be prepared should more virulent flu strains emerge. The agency also coordinated planning with the nuclear industry, with the goal of ensuring that the nuclear sector is prepared to address the challenges of a pandemic and maintain the standards of safety and security required for operations. The NRC will continue to update its pandemic plan whenever possible to take advantage of improving communications and other technology.

International Activities

The NRC's international responsibilities include participation in activities that support U.S. Government compliance with international treaties and agreements; export and import licensing of nuclear facilities, equipment, and materials; programs of bilateral nuclear cooperation and assistance; and multinational nuclear safety organizations such as IAEA and the Organisation for Economic Co-operation and Development's Nuclear Energy Agency (NEA). The agency is also the U.S. representative to the IAEA's radiation, waste, and transportation safety standards committees and NEA's technical standing committee.

Export and Import Licensing

The NRC issued a final rule updating 10 CFR 110, "Export and Import of Nuclear Equipment and Material," that revised the definition of radioactive waste, incorporated changes to Appendix P, "Category 1 and 2 Radioactive Material," and rewrote and clarified 10 CFR 110.23 to 10 CFR Part 110, "General License for the Export of Byproduct Material."

The NRC completed reviews for, and issued as appropriate, 85 import/export authorizations within 60 days of receipt of applications in FY 2010. The NRC's import/export licensing reviews ensure that nuclear equipment and material are transported and used in a manner consistent with applicable U.S. law and international requirements.

Bilateral Cooperation and Assistance

In FY 2010, the NRC took the first steps in preparing an information exchange arrangement with the United Arab Emirates and in expanding cooperation with China (on the design, construction, and operation of first-of-a-kind Westinghouse AP1000 nuclear power plants in China). The agency expanded cooperation with Vietnam on establishing siting requirements for potential nuclear power plant construction in that country. The NRC also began cooperation with Thailand and Indonesia on establishing the basic regulatory infrastructure needed for oversight of a nuclear power program. The agency also engaged in bilateral inspection training activities with Finland and China, which are building new reactors.

Multilateral Nuclear Safety Organizations

The NRC continued to work with IAEA in FY 2010 to revise TS-R-1, "Regulations for the Safe Transport of Radioactive Materials," to adopt a U.S.-based approach to fissile material exceptions, dose-based exemptions for naturally occurring radioactive materials, and transitional arrangements based primarily on package service lifetime. Additionally, the agency participated in an IAEA-coordinated research project on spent fuel performance assessment and research, which is focused on a wide range of issues dealing with wet and dry storage of spent nuclear fuel. Of particular interest are those issues that could affect the licensing of extended dry storage of high burnup fuel. U.S. participation seeks to leverage international experiences, as the Nation's policy on spent fuel management will likely include storing spent fuel in dry cask storage systems for extended periods.

The NRC is engaged both domestically and internationally in efforts to enhance nuclear safety and security through the regulatory oversight of radioactive sources. In May 2010, the NRC attended an IAEA open-ended meeting of technical and legal experts to share information on States' implementation of IAEA's Code of Conduct for the Safety and Security of Radioactive Sources. In addition, the agency continued radioactive source-related assistance to the countries of the Commonwealth of Independent States, expanded a provision of radioactive source-related assistance to include selected countries of Africa, Latin America, and Southeast Asia, conducted regional workshops on the physical protection of radioactive sources, and continued coordination with source-related assistance provided by the IAEA and others. The agency also worked with other U.S. Government agencies, such as the Departments of State, Energy, and Commerce, and the National Security Council, and with IAEA to develop international security guidance documents for radioactive sources.

The NRC continued its support of an effective international safeguards and nonproliferation regime. The agency participated in the U.S. Support Program to identify funding and support for IAEA safeguards and counterproliferation programs in FY 2010. The NRC also supported the U.S. Government's participation in the Nuclear Non-Proliferation Treaty review conference. The NRC participated in several consultancy meetings tasked to prepare an IAEA guidance document on how to develop and implement a material control and accountancy and security program within a country. In addition, the agency contributed to generic IAEA guidance for uranium enrichment facilities. IAEA's main guidance on nuclear material control and accountancy for nuclear security at facilities has been drafted and distributed for internal review by different participating agencies.

The NRC continues to support the development and implementation of programs to leverage the knowledge and resources within the international regulatory community in the licensing of new reactor designs. The agency continued its leadership role in the Multilateral Design Evaluation Program (MDEP), through which regulatory authorities in 10 countries share expertise and resources in reviewing new reactor designs. Currently, the program consists of three issue-specific and two design-specific working groups. Led by the United States, the Digital Instrumentation and Controls Working Group, drafted common positions in digital instrumentation and controls system design. The Vendor Inspection Cooperation Working Group conducted several parallel inspections that involved more than one regulator, and the Codes and Standards Working Group is nearing completion of a project to compare the pressure boundary codes of five member countries. The design-specific working groups, based on the Westinghouse AP1000 and the AREVA evolutionary power reactor designs, also established subworking groups. In FY 2010, the Policy Group, which is the governing body of the program, began modifying the MDEP terms of reference to establish a process for additional countries to join.

STRATEGIC GOAL 2: SECURITY

Ensure Adequate Protection in the Secure Use and Management of Radioactive Materials

Strategic Outcome

The NRC's strategic outcome associated with its goal to ensure adequate protection in the secure use and management of radioactive materials is the following:

- Prevent any instances where licensed radioactive materials are used domestically in a manner hostile to the security of the United States.

RESULTS: In FY 2010, the NRC achieved its Security goal strategic outcome.

Performance Measures

The NRC uses performance measures to assess whether the agency has met its Security goal. Performance measures are set at a different risk level than the strategic outcomes, and missing a performance measure signals that safety levels may have deteriorated at the agency strategic planning level. If the NRC misses a performance measure, the agency will take corrective actions to bring the measure back into the target range. Table 2 shows the agency's annual performance measures and their outcomes for the past 6 years.

The NRC met all of the FY 2010 performance measure targets for its Security goal.

Table 2

FY 2010 SECURITY GOAL PERFORMANCE MEASURES

Performance Measure	2005	2006	2007	2008	2009	2010
1. Number of unrecovered losses or thefts of risk-significant radioactive sources is zero.	0	0	0	0	0	0
2. Number of substantiated cases of theft or diversion of licensed, risk-significant radioactive sources or formula quantities of special nuclear material, or attacks that result in radiological sabotage, is zero.	0	0	0	0	0	0
3. Number of substantiated losses of formula quantities of special nuclear material or substantiated inventory discrepancies of formula quantities of special nuclear material that are judged to be caused by theft or diversion, or by substantial breakdown of the accountability system, is zero.	0	0	0	0	0	0
4. Number of substantial breakdowns of physical security or material control (i.e., access control containment or accountability systems) that significantly weaken the protection against theft, diversion, or sabotage is less than or equal to one.	0	0	0	0	0	0
5. Number of significant unauthorized disclosures of classified or safeguards information is zero.	0	0	0	0	0	0

Analysis of FY 2010 Results

1. **Unrecovered losses or thefts:** This measure tracks any loss or theft of radioactive nuclear sources that the NRC has determined to be of significant risk. The measure tracks the agency's performance in ensuring that licensees properly account for radioactive sources of significant. The ability to account for these sources is vital to securing the Nation's critical infrastructure from dirty bomb attacks or other radiological crimes. There was no loss or theft of radioactive nuclear material that the NRC determined to be risk significant during FY 2010.

2. **Thefts or diversion:** This measure tracks whether NRC-licensed facilities maintain adequate protective capabilities to prevent theft or diversion of nuclear material or sabotage that could result in substantial harm to public health and safety. There were no substantiated cases of theft or diversion of licensed, risk-significant radioactive sources or formula quantities of special nuclear material or attacks that resulted in radiological sabotage during FY 2010.

3. **Loss or inventory discrepancy:** This measure tracks whether special nuclear material is accounted for and that losses of this material do not occur that could lead to the creation of an improvised nuclear device or other type of nuclear device. The measure also tracks whether the systems in place at NRC-licensed facilities maintain accurate inventories of the special nuclear material that the facilities process, use, or store. There were no substantiated losses of formula quantities of special nuclear material or substantiated inventory discrepancies of formula quantities of special nuclear material that were caused by theft or diversion or by substantial breakdown of the accountability system during FY 2010.

4. **Substantial breakdowns of physical security:** This measure tracks any breakdowns in access control, containment, or accountability systems that significantly weakened the protection against theft, diversion, or sabotage for nuclear materials

the agency has determined to be of significant risk. There were no substantial breakdowns of physical security during FY 2010.

5. **Significant unauthorized disclosures:** This measure includes significant unauthorized disclosures of classified or safeguards information that cause damage to national security or public safety. This measure tracks whether information that can harm national security (classified information) or cause damage to the public health and safety (Safeguards Information) has been stored and used in such a way as to prevent its disclosure to terrorist organizations, other nations, personnel without a need to know, or the public. There were no significant disclosures that caused damage to national security or public safety during FY 2010.

Nuclear Security Programs

The NRC must remain vigilant to protect the security of nuclear facilities and materials. The agency achieves its common defense and Security goal with licensing and oversight programs similar to those employed in achieving its Safety goal. The aim is to allow licensees to realize the benefits of nuclear materials through their secure use while placing only necessary regulatory requirements on licensees.

New and Operating Reactor Security

The NRC conducts a robust security inspection program within the Security Cornerstone of the agency's Reactor Oversight Process. The Security Cornerstone focuses on five key attributes of licensee performance: access authorization, access control, physical protection systems, material control and accounting, and response to contingency events. Through the results obtained from all oversight activities, including baseline security inspections and performance indicators, the agency determines whether licensees are in compliance with NRC requirements and can provide high assurance of adequate protection against the design-basis threat for radiological sabotage. There were no substantial breakdowns of physical security at any commercial nuclear power plant in FY 2010.

The NRC regularly carries out force-on-force inspections at least once every 3 years at each commercial operating nuclear power plant as part of its comprehensive security program. The agency uses these inspections to evaluate the effectiveness of security programs to prevent radiological sabotage.

Force-on-force inspections assess the ability of nuclear facilities to defend against the applicable design-basis threat, which characterizes the adversary against which licensees must design appropriate defenses, such as physical protection systems and response strategies. A force-on-force inspection includes tabletop drills and simulated combat between a mock commando-type adversary force and the site security force. During the attack, the adversary force attempts to reach and damage key safety systems and components at a nuclear power plant, steal material at a Category I fuel facility, or gain control of safeguarded material. In FY 2010, the agency completed 26 force-on-force inspections at nuclear power plants and one force-on-force inspection at a Category I fuel facility and submitted its fifth annual report to Congress on the results of the security inspection program.

In March 2009, the NRC issued "Power Reactor Security Requirements," revising and creating several security regulations under 10 CFR Parts 50, 52, 72, and 73. The full compliance date for 10 CFR Part 73, "Physical Protection of Plants and Materials," was March 31, 2010. Licensees are required to meet more than 280 areas of compliance. Forty licensees submitted requests for exemption in accordance with 10 CFR 73.5, "Specific Exemptions," for exemption from the compliance date for specific parts of the regulation, with the intention of meeting all other requirements by the full implementation date. The exemption requests varied significantly in the amount of time needed to be in full compliance with the new regulations, based on individual licensee security upgrades. The agency processed all 40 requests in a timely manner and granted appropriate relief while ensuring adequate security.

The NRC also enhanced its allegation and inspection programs based on a lessons-learned review that followed an agency investigation into reports of inattentive

security officers at the Peach Bottom nuclear power plant in Pennsylvania. To address lessons learned, on February 2, 2010, the agency finalized guidance in the areas of contacting those who make allegations, engaging licensees with requests for information, and independent validation of licensee inputs, among other enhancements.

The NRC continued developing a rulemaking on access authorization at power reactors under construction and the regulatory guidance needed to support this rulemaking. The intent of the rule is to deter and detect malicious acts during construction that could later be exploited and interfere with safety- and security-related structures, systems, or components after the plant becomes operational.

Spent Fuel, Fuel Cycle Facility, and Transportation Security

The NRC completed its FY 2010 core security inspection program at NRC-licensed materials and waste facilities and fuel cycle facilities. It also completed six site visits to review licensee implementation of the security orders for independent spent fuel storage installations.

In FY 2010, the NRC continued efforts to establish and monitor classified information security programs for uranium enrichment vendors and mixed-oxide facilities, including readiness reviews at multiple fuel cycle facilities. These reviews included evaluation of physical and information system security at these sites, licensee contractors performing classified work, and foreign ownership, control, or influence considerations in support of the facility clearance. In addition, NRC personnel participated in Quadripartite Working Group and DOE meetings on the protection of sensitive information associated with the URENCO USA enrichment facility.

The NRC regularly carries out force-on-force inspections at Category I fuel facilities as part of its comprehensive security program. The agency uses these inspections to evaluate the effectiveness of security programs to prevent radiological sabotage and the theft or diversion of Category I material. The agency conducts force-on-force inspections at least once every 3 years at each Category I fuel facility.

The NRC continued security rulemaking activities to stabilize its security requirements for licensees. The agency published a proposed rule that would add a new 10 CFR Part 37, "Physical Protection of Byproduct Material," and made conforming changes to other parts of the regulations. The proposed rule will put in place generally applicable requirements for licensees that possess Category 1 and Category 2 radioactive materials, as defined by the IAEA Code of Conduct on the Safety and Security of Radioactive Sources. The proposed rule addresses physical protection at the facilities during transit, as well as access to materials. The agency also developed draft technical bases to support the commencement of a rulemaking in FY 2011 about physical protection requirements for fuel cycle licensees and spent fuel cask certificate holders, and a separate rulemaking on security requirements for independent spent fuel storage installations.

Nuclear Material Users

The NRC continued its efforts to mitigate the potential risk of terrorist threats through enhanced security and controls for the use, storage, and transportation of risk-significant byproduct material and spent nuclear fuel. In collaboration with DHS, DOE, and other Federal, State, and local agencies, the NRC continued to assess the potential use of risk-significant sources in radiological dispersal devices and to coordinate efforts to enhance radioactive source protection and security. The NRC also worked with Agreement States to implement requirements for licensees that enhance the security and control of risk-significant radioactive material, including development of an inspection program to verify the implementation of these measures.

The Energy Policy Act of 2005 established an interagency task force on radiation source protection and security, led by the NRC, to evaluate and provide recommendations to the President and Congress on the security of radiation sources in the United

States from potential terrorist threats, including acts of sabotage, theft, or use of a radiation source in a radiological dispersal device or radiological exposure device. In FY 2010, the NRC staff participated in several subgroups that developed the 2010 Chairman's Task Force Report. The agency provided the report to the President and Congress in August 2010.

The NRC staff participated in activities related to the Government Coordinating Council, which enables interagency and cross-jurisdictional coordination on critical infrastructure and key resources, including transportation and material security. The staff also participated in trilateral meetings with DHS and DOE's National Nuclear Security Administration to enable coordination among the participants on issues related to radioactive material security.

The Commission is reviewing a final rule on generally licensed device restriction. If approved, this rulemaking would limit the allowable quantity of radioactive material in generally licensed devices.

Control of Radioactive Sources

Both the NRC and Agreement States implemented measures, via orders and other regulatory binding requirements, to put into practice requirements imposed on licensees that enhance the security and control of risk-significant quantities of radioactive material. In FY 2010, the agency completed work on development of the proposed new 10 CFR Part 37, which captures these requirements, lessons learned during implementation, and other factors. The objective of the proposed rulemaking is to ensure that effective security measures are in place for the protection of IAEA Category 1 and Category 2 quantities of radioactive material against the possibility of its dispersion for malevolent purposes.

The NRC also implemented the National Source Tracking Rule, which requires licensees to report information on the possession of IAEA Category 1 and 2 radioactive sources (i.e., nationally tracked sources). The rule requires NRC and Agreement State licensees to report transactions involving the manufacture, transfer, receipt, disassembly, and disposal of nationally tracked sources. In FY 2010, licensees completed the first annual inventory reconciliation of their nationally tracked sources.

The National Source Tracking System, and the future Web–based Licensing System and License Verification System, are key components of a comprehensive program for the security and control of radioactive material. The NRC plans to integrate all three systems into a common system environment and architecture to form an integrated source management system that will include information on all U.S. licensees and more than 70,000 risk-significant radioactive sources possessed by approximately 1,400 licensees. The integrated system will provide licensees, regulators, and Federal agencies with an additional round-the-clock means of determining the legitimacy of individuals possessing or seeking to obtain radioactive material—to ensure that the materials are obtained only in authorized amounts by legitimate users.

The NRC completed construction of a secure enclave within the Operations Center to enhance the response capability for security events and to support the NRC's transition to DHS's Homeland Secure Data Network for classified information. The new enclave provides improved security for the processing of Safeguards Information and classified information. The agency is also using this enclave to prototype data presentation technology and workstation layouts in preparation for the Operations Center's move to a new building in late 2012.

International Security

The NRC continued its significant participation in implementing portions of the IAEA Code of Conduct on the Safety and Security of Radioactive Sources, as well as its participation in IAEA committees that are developing guidance documents for the security of radioactive sources during use, storage, and transport. The agency's involvement in these committees enhances security and public safety and contributes to international and domestic regulatory consistency. During FY 2010, the agency issued 175 licenses for the export or import of Category 1 and Category 2 radioactive materials as defined by the code.

Integrated and Coordinated Security Activities

The NRC has developed and enhanced working relationships with the Federal Bureau of Investigation (FBI), DHS, Nuclear Energy Institute (NEI), power reactor licensees, and State and local law enforcement agencies to create integrated approaches to security within the nuclear sector. One significant outcome is the Integrated Pilot Comprehensive Exercise (IPCE). The IPCE is a voluntary, collaborative effort led by the FBI with the support of DHS, the NRC, and NEI. The IPCE incorporates Federal, State, and local law enforcement tactical response planning and operations into the concept of integrated response by providing law enforcement tactical teams with opportunities to prepare for and respond to simulated security incidents inside commercial nuclear power plants. An IPCE was conducted in July 2010.

The NRC participated in many other nuclear sector activities under DHS's National Infrastructure Protection Plan framework, such as the Government Coordinating Council, Critical Infrastructure Partnership Advisory Council, Federal Senior Leadership Council, and Research and Development Working Group. The NRC also contributed to national policy documents, including the Nuclear Sector-Specific Plan, Nuclear Sector Critical Infrastructure and Key Resources Protection Annual Report, and the National Critical Infrastructure and Key Resources Annual Report.

Cyber Security

The NRC issued 10 CFR 73.54, "Protection of Digital Computer and Communication Systems and Networks," in March 2009. Licensees and COL applicants are required to provide high assurance that nuclear power plant safety, security, and emergency preparedness (SSEP) functions are adequately protected from cyber attacks up to and including the design-basis threat. This new regulation required operating power reactor licensees to submit a cyber security plan, including a proposed implementation schedule, to the NRC no later than November 23, 2009. The staff reviewed licensees' proposed schedules to fully implement the programmatic requirements of 10 CFR 73.54 during FY 2010.

In January 2010, the NRC issued Regulatory Guide 5.71, "Cyber Security Programs for Nuclear Facilities," which describes an acceptable method for complying with the agency's regulations about the protection of digital computers, communications systems, and networks from cyber attacks in support of 10 CFR 73.54. This guide is based on standards from the National Institute of Standards and Technology, DHS, and the Institute of Electrical and Electronics Engineers and is tailored to address the specific needs of new and existing plant systems performing or supporting SSEP functions.

In 2009, consistent with its statutory authority, the Federal Energy Regulatory Commission issued Order 706B relating to cyber security requirements at commercial nuclear power plants. In December 2009, the NRC signed a memorandum of understanding with the North American Electric Reliability Corporation, which is overseen by the Federal Energy Regulatory Commission, clarifying the regulatory roles and responsibilities of each organization, including inspection protocols and enforcement actions. The memorandum of understanding clarified that the NRC is responsible for inspecting digital assets that can have an adverse impact on SSEP functions. The North American Electric Reliability Corporation is responsible for inspecting digital assets that can affect the continuity of electric power generation, and for enforcing compliance with its Critical Infrastructure Protection Program reliability standards.

Costing to Goals

The NRC is working to improve its cost management capabilities to better align its costs with desired outcomes. This year's Performance and Accountability Report presents the full cost of achieving the safety and security goals for the agency's programs, Nuclear Reactor Safety and Security and Nuclear Materials Safety and Security. The cost of achieving the agency's safety goal was $1,074.8 million, and the cost of achieving the agency's security goal was $65.7 million (see Figure 20).

Figure 20
NRC SAFETY AND SECURITY COSTS (In Millions)

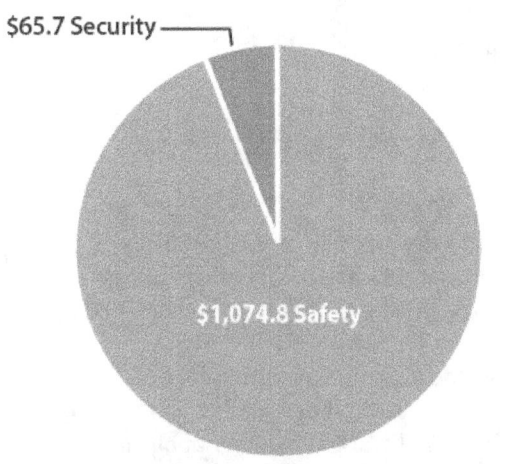

$65.7 Security

$1,074.8 Safety

Organizational Excellence Objectives

The NRC has three Organizational Excellence Objectives: openness, effectiveness, and operational excellence. These objectives are critical components to carrying out the agency's regulatory mandate to serve the American people.

Openness

The openness objective explicitly recognizes that the public must be informed about, and have a reasonable opportunity to participate in, the NRC's regulatory processes. The NRC is firmly committed to transparency, participation, and collaboration as key principles governing the agency's relationship with the public and other stakeholders. The agency has demonstrated its commitment to these openness principles through its long-standing efforts to keep stakeholders informed and involved in the NRC's regulatory process. The NRC's response to the Open Government Directive reaffirms that commitment, extending agency efforts through the use of social media, such as citizen-engagement tools that quickly gather and rank stakeholder ideas, and collaboration technologies such as Web conferencing tools that broaden participation in public meetings. The NRC published its Open Government Web page on February 4, 2010, at http://www.nrc.gov/open. In April 2010, the agency published its Open Government Plan that serves as a public roadmap to how the agency will incorporate the principles of open government into its core mission objectives.

Nuclear Reactor Safety

Operating Reactors

The NRC held many public meetings during FY 2010 seeking public input on updates to the GEIS and NUREG-1801, "Generic Aging Lessons Learned (GALL) Report." Among the benefits of these meetings was to allow the public to identify significant environmental issues in the proposed operation of Watts Bar Unit 2. The topics of other public meetings included fire protection. Diablo Canyon license renewal, B&W Medical Isotope Production Systems, and the shortage of medical isotope Molybdenum-99.

The NRC also held monthly public meetings during FY 2010 to discuss the Reactor Oversight Process. Participants discussed suggestions for improvement, questions, and program implementation issues. Additionally, the agency continued to provide accurate and timely information to the public by ensuring that nonsensitive, unclassified regulatory documents are released to the public by the sixth working day after the document date. The NRC routinely holds public meetings to present the agency's assessments of safety performance at nuclear reactor sites.

The NRC maintains information on license renewal for commercial operating power reactors on its Web site. Processes, regulations, and inspection reports for the Reactor Oversight Process are also available on the NRC Web site (http://www.nrc.gov/NRR/OVERSIGHT/ASSESS)

New Reactors

The NRC updated project status and schedules for new reactor licensing activities monthly during FY 2010 and made them available on the NRC Web site (http://www.nrc.gov/reactors/new-reactors.html). The NRC Web site received approximately 50,000 hits per month for information on new reactor licensing activities.

The NRC held more than 140 public meetings on new reactor activities in FY 2010. These meetings engage stakeholders in the regulatory process, provide information on public participation in the environmental review process, solicit comments on the scope of environmental impact statements, and provide information on lessons learned about locating sites and environmental reviews.

The NRC also held many public meetings to provide a forum for stakeholders to participate and comment on staff proposals for the closure of ITAAC and on licensee assessment and enforcement topics. The agency conducted many activities to support implementation of construction inspection. For example, the agency conducted a workshop on vendor oversight and new reactor construction attended by more than 600 public participants who discussed topics of mutual interest, including counterfeit, fraudulent, and substandard items, safety culture, and the American Society of Mechanical Engineers survey process. The agency also held public workshops on proposed rulemaking activities for design certification rule templates, ITAAC maintenance, and access authorization and physical protection during new reactor construction.

Nuclear Materials and Waste Safety

The NRC continued its active participation in many meetings to inform the public about its activities. Agency representatives attended meetings for the Institute of Nuclear Materials Management Spent Fuel Seminar, regional meetings of the Council of State Governments, the U.S. Transport Council, and the NEI Dry Cask Storage Forum on radioactive material transportation and spent fuel storage matters.

In its continuing efforts to reach out to stakeholders, the NRC conducted its fifth annual Fuel Cycle Information Exchange conference in July 2010. The Fuel Cycle Information Exchange addresses a broad range of issues in the licensing and oversight of new and operating fuel facilities and potential developments for future reactors and fuel cycles. It provides a forum for presentations and panel discussions involving regulators, industry, and public stakeholders, both domestic and international.

The NRC also met with stakeholders to discuss spent fuel reprocessing issues. Agency representatives met with NEI representatives in May 2010 to discuss the technical basis to resolve several gaps in the identified regulations related to reprocessing. A more comprehensive workshop that provided the perspectives of industry and members of the public took place in September 2010.

The Commission directed staff to make modest enhancements to the fuel cycle oversight process to enhance its effectiveness and efficiency, such as providing licensees credit for a corrective action program. The Commission also directed staff to prepare a paper comparing the integrated safety analysis for fuel cycle facilities and probabilistic risk assessment of reactors.

The NRC also met with stakeholders on many occasions to discuss how safety culture could be applied to the materials program and the best way to implement a safety culture policy. The NRC held stakeholder meetings on the proposed new 10 CFR Part 37, which codifies security requirements into NRC regulations.

In FY 2010, the NRC held public meetings to discuss the decommissioning plan and proposed activities at the Hematite site and with the uranium recovery industry to provide information to facilitate the preparation of license application submittals. The agency held 17 technical meetings with decommissioning licensees and uranium recovery facility applicants and licensees that were open to the public. In June 2010, the NRC staff held a briefing for the Commissioners on the issue of blending low-level waste, in which representatives from other Federal and State regulatory agencies, interested stakeholders, and Native American Tribal Governments were invited to make presentations and discuss their concerns.

Effectiveness

The drive to improve performance in government, coupled with increasing demands on the NRC's resources, requires the NRC to become more effective, efficient, and timely in its regulatory activities. The NRC's effectiveness initiatives sharpen the agency's

focus on safety and security and ensure that its available resources are optimally directed toward accomplishing the agency's mission.

Nuclear Reactor Safety

Operating Reactors

In FY 2010, the NRC implemented the Risk Tools Enhancement Program, which coordinates the continual improvement of the many risk tools used by staff across the agency in order to promote the development of high-quality tools and ensure their efficient and effective use by NRC staff and management. The agency continues to rely and build on industry operating experience and available IT to improve its programs (for example, NRC efforts to optimize the Reactor Oversight Process). The agency updated the infrastructure for the license renewal program, which included updating the GALL report and the GEIS report for license renewal to increase the efficiency of the program.

In its efforts to improve regulatory programs, the NRC published the final rule on alternate fracture toughness requirements for protection against pressurized thermal shock events. The rule increases realism of calculations used to examine the pressurized water reactor susceptibility to a phenomenon known as pressurized thermal shock.

The NRC published Regulatory Guide 1.189, Revision 2, "Fire Protection for Nuclear Power Plants," in October 2009 with important exceptions, and clarifications to NEI's guidance document, NEI 00-01, Revision 2, "Guidance for Post Fire Safe Shutdown Analysis," issued May 2009. The agency also provided guidance on fire-induced circuit failures in Regulatory Guide 1.205, Revision 1, "Risk-Informed, Performance-Based Fire Protection for Existing Light-Water Nuclear Power Plants," issued December 2009, with important exceptions and clarifications to NEI 04 02, Revision 2, "Guidance for Implementing a Risk-Informed, issued April 2008, Performance-Based Fire Protection Program," on the transition to a risk-informed, performance-based fire protection program based on National Fire Protection Association Standard 805.

New Reactors

For the new reactor license applications currently under review, the NRC used earned value management project health indicators to determine overall project health, and improve schedule compliance, resource use, and improve the efficiency of the project under review. As a result of implementing earned value management, the agency increased the effectiveness of new reactor licensing in three ways. First, it focused limited resources on the new reactor projects that are expected to complete licensing and construction and begin operation in the near term. Second, it identified and minimized risks to project schedules and review completions. Third, it managed resource use across many complex new reactor licensing applications.

Nuclear Materials and Waste Safety

The NRC developed a plan to address ongoing revisions to the national strategy for ensuring public health and safety and environmental protection in managing spent nuclear fuel and high-level waste. The plan integrates spent nuclear fuel regulatory activities to address more effectively the regulatory and licensing aspects of extended storage and transportation (i.e., greater than 120 years), reprocessing, and disposal of spent nuclear fuel and high-level waste. The purpose of the plan is to ensure that the regulation of the back end of the fuel cycle accomplishes safety, security, and environmental protection in an efficient and effective manner and that decisions made about one component or area of this system adequately consider other components or areas (i.e. treating spent fuel and high-level waste regulation as a system of interrelated activities). By coordinating the approach for regulation of spent nuclear fuel and high-level waste storage, potential reprocessing, transportation, and disposal, the agency can improve the efficiency and effectiveness of NRC regulatory processes and provide stability and predictability for stakeholders in a dynamic environment.

As part of the NRC's license review process, the agency performs an acceptance review to determine if the license application contains adequate information. To aid the review of uranium recovery in-situ leach applications, the NRC published a GEIS to be used in the agency's application review process. The GEIS will improve the review process and ensure that the NRC staff does not expend resources reviewing submissions that contain incomplete or inadequate information. The agency estimates that the GEIS will save as much as $7 million total for all application reviews and reduce review time by 2 years per application. In FY 2010, the NRC signed a memorandum of understanding with the U.S. Bureau of Land Management to allow for cooperation between the NRC and the Bureau on environmental review documents to meet the National Environmental Policy Act requirements.

Operational Excellence

This objective focuses on activities related to financial management, the management of human capital, information management, and infrastructure management. This objective supports the NRC by ensuring that the necessary corporate support is in place to help the agency achieve its mission.

Financial Management

The NRC made substantial progress in modernizing its financial systems in FY 2010. An e-Travel system, deployed in FY 2009, was expanded in FY 2010 to include split payments (payment of a portion to the traveler's credit card and a portion to a bank account) and foreign travel. The agency transitioned to a new core accounting system, deployed in October 2010. The new accounting system lays the foundation for significantly enhanced performance by providing up-to-date query and reporting tools and the ability for NRC users to easily navigate across modules to obtain and analyze business information. The system will provide staff with access to real-time data on demand.

The NRC continued to achieve operational excellence in financial reporting. It received an unqualified opinion on its FY 2009 financial statement, with no material weaknesses. In addition, it received its ninth consecutive Certificate of Excellence in Accountability Reporting award from the Association of Government Accountants.

During FY 2010, the NRC increased its emphasis on improvements that align the spending forecast to the agency's acquisition planning. The Chief Financial Officer led an agencywide effort to significantly reduce previous fiscal year unliquidated obligations. Increased emphasis was also placed on improving service and outreach to internal stakeholders.

Management of Human Capital

There continues to be a shortage of personnel in the nuclear sector as the current workforce retires and normal attrition occurs. The NRC has a program to provide grants to educational institutions in the areas of curriculum development, faculty development, fellowships, and scholarships to 4-year institutions, trade schools, and community colleges. In FY 2010, the NRC made 100 grants to educational institutions in 33 States. These grants, focused on nuclear engineering, health physics, radiochemistry, and other related areas that benefit the nuclear sector, help expand the workforce in nuclear safety and nuclear-related disciplines and develop the next-generation nuclear workforce.

The Federal Employee Viewpoint Survey was conducted in FY 2010 and the NRC's results were excellent. For the first time, the NRC was ranked first in all of the Human Capital Indexes. The key to NRC's success on this survey has been the management's unwavering commitment to continue to improve human capital programs and communications. Agency employees believe that the results of the Federal Employee Viewpoint survey will make a difference.

To follow-up on the Office of the Inspector General's 2009 Safety Culture and Climate Survey, the NRC has implemented a number of programs for continuous improvement. These include placing more emphasis

on (1) Open, Collaborative Working Environment communications and programs, as well as emphasizing all employees' connections to the mission, including corporate and support offices; (2) knowledge management strategies; (3) placing more emphasis on "staying connected" for those offices remotely located at headquarters; (4) looking for ways to enhance internal communication mechanisms; and (5) identifying and implementing more effective tools to improve performance management and feedback.

Forty-five percent of NRC employees at the end of FY 2010 had 5 or fewer years experience at the NRC. Approximately 17 percent of the employees are eligible for retirement in FY 2010. The NRC continues to enhance its Knowledge Management Program by expanding the agency's Knowledge Center, an agencywide collection of electronic communities of practice that enables staff to collaborate, capture, and share knowledge. During FY 2010, the agency conducted its first agencywide Knowledge Management Fair in conjunction with its 35th anniversary celebration.

The NRC significantly upgraded its learning management system and continues to leverage its use for efficient training delivery. The learning management system has returned substantial cost savings through time compression and avoided travel.

In accordance with the E-Government Initiative, the NRC adopted electronic official personnel folders (e-OPF) and trained staff on using the new system. E-OPF allows employees real-time access to their personnel records and eliminates the need for paper copies.

The NRC kicked off the Veterans' Hiring Initiative to promote and enhance employment opportunities for veterans, as outlined in the President's Executive Order 13518, "Employment of Veterans in the Federal Government," dated November 9, 2009. The agency has developed an operational plan that lays out the activities that it will undertake to maintain and demonstrate its commitment to veterans' employment through discrete goals and actions.

Infrastructure Management

The NRC broke ground during FY 2010 on a new building that will house at least 1,300 NRC employees. The agency currently has staff at four interim locations in addition to the Headquarters One and Two White Flint buildings in Rockville, MD. The agency established the Staying Connected Working Group to maintain the feeling of employee cohesiveness to compensate for the dispersion of agency personnel. During FY 2010, suggestions from the working group resulted in a number of improvements, including increased shuttle frequency and the broadcast of all agency events via video teleconference. Additionally, the agency maintains touchdown stations in each Headquarters location equipped with telephones and computers to enable employees to conduct business while at locations other than their primary duty station.

The NRC enhanced employee security and safety with a Physical Access Control System upgrade at all NRC facilities. This project coincided with the issuance this year of new Homeland Security Presidential Directive 12 "Policies for a Common Identification Standard for Federal Employees and Contractor," dated August 27, 2004, to all staff that was also completed this year. The upgrade will shift the NRC's security strategy from an emphasis on interior-based controls to perimeter-based access for vehicles and pedestrians.

The NRC supported its open government plan www.nrc.gov/open by expanding the Web streaming program at Headquarters from 50 to 100 meetings of significant public interest during FY 2010. The expanded program increases transparency, participation, and collaboration with the public.

Information Technology and Management

The NRC continued to partner with internal stakeholders to identify opportunities to improve program performance and information availability through the use of IT solutions. Progress continued in several major focus areas to achieve operational excellence through more effective information management, effective IT infrastructure, and continuous customer service improvements.

Effective information management provides NRC staff and stakeholders with effective access to the information they need to fulfill the agency's mission. FY 2010 accomplishments in this area included analyses for reducing the number of system sign-ons; improvements to the NRC's SharePoint program; development and deployment of a new Open Government Web site; many activities for the modernization of the Agencywide Documents Access and Management System (ADAMS); and issuance of an open government plan that scored very high as a strong, initial blueprint to increase transparency, participation, and collaboration.

Effective IT infrastructure ensures that the NRC has a reliable and responsive foundation of technology to support business needs and agency operations. Accomplishments in this area primarily focused around two broad themes—working from anywhere and working with anyone—and included implementation of a "secure laptop loaner pilot" providing an increased number of laptops for mobile users across the agency, modernization of remote access systems for telecommuters and resident inspectors, deployment of Internet Explorer 8 to the Enterprise, deployment of Network Access Control to enhance controls on the NRC network, transition from WITS 2001 to WITS 2003, and an upgrade to Microsoft Office 2007 to ensure a standardized suite of office products.

Another primary focus area is service, a key component of operational excellence across the agency. The NRC solicited feedback from employees through an IT survey. Survey results and followup actions were posted for all NRC employees; this is an ongoing activity. The agency has developed communication plans to further improve service levels and expectations. An IT Service Catalog listing IT services was developed and provided electronically to improve service request capabilities.

Program Evaluations

The NRC conducted a number of program evaluations of its regulatory operations during FY 2010. The evaluations were conducted for both the nuclear reactor and the nuclear materials programs.

Operator Licensing Program

Before the NRC licenses an individual to operate or supervise the controls of a commercial nuclear power reactor, the applicant must complete extensive training and pass rigorous examinations. Once licensed, operators and senior operators must comply with a number of requirements to maintain and renew their licenses. In FY 2010, an agency review team evaluated the operator licensing programs of two regions for their overall effectiveness and adherence to the guidance contained in NUREG-1021, Revision 9, "Operator Licensing Examination Standards for Power Reactors," issued July 2004, and other policy documents. The operator licensing programs are broken down into seven functional areas that are rated as either "satisfactory," or "needs improvement." The review team found the operator licensing programs in the two regions to be in accordance with the examination standards and assessed all areas as satisfactory. The review team also commended the regions' efforts to improve the quality of their examination packages.

Reactor Oversight Process

The NRC completed a self-assessment of the Reactor Oversight Process in April 2010. SECY-10-0042 entitled, "Reactor Oversight Process Self-Assessment for Calendar Year 2010" is available on the NRC Web site). The results of the calendar year 2010 self-assessment indicated that the Reactor Oversight Process met its program goals and achieved its intended outcomes. The assessment found the Reactor Oversight Process to be objective, risk-informed, understandable, and predictable, and it met the agency goals of ensuring safety, openness, and effectiveness. The NRC maintained its focus on stakeholder involvement and continued to improve the

Reactor Oversight Process. The agency implemented improvements to address issues that were raised internally, recommended by independent reviews, and obtained from internal and external stakeholder feedback.

The NRC inspection and assessment program independently verified that nuclear power plants were operated safely and securely. The NRC revised the assessment program to incorporate lessons learned from implementation of the safety culture enhancements and continued to ensure that the staff and licensees acted as necessary to address identified performance issues. The agency continues to improve the performance indicator program to ensure that the performance indicators are meaningful inputs to the Reactor Oversight Process, and it actively solicits input from internal and external stakeholders to further improve the Reactor Oversight Process based on stakeholder feedback and lessons learned.

Integrated Materials Performance Evaluation Program Reviews of Selected Agreement States and NRC Regional Offices

The NRC evaluates its own regional materials programs and Agreement State radiation control programs using performance indicators to ensure that public health and safety is adequately protected. With the assistance of the Agreement States, the NRC completed nine Integrated Materials Performance Evaluation Program reviews to determine the adequacy and compatibility of the programs in the evaluated Agreement States, one review of the materials licensing and inspection program in NRC Region I, and one review of the sealed source and device evaluation program at NRC Headquarters during FY 2010. Region I was found satisfactory (the highest level) for all areas of the review; there were no recommendations. The Headquarters' sealed source and device program was found satisfactory for all areas of the review. The review team made one recommendation to the Headquarters' sealed source and device program to ensure that the appropriate documentation is tied to sealed source and device registries for enforceability. Headquarters addressed the recommendation by implementing increased quality control in documentation. For the nine Agreement State reviews conducted in FY 2010, seven of the Agreement States were found to be adequate and compatible (the highest finding) and two of the Agreement States were found to be adequate but needs improvement, and compatible.

Fuel Cycle Licensing and Inspection Program

The NRC's Fuel Facilities Licensing and Inspection Program regulates the Nation's nondefense-related fuel fabrication facilities. The licensing program issues licenses to facilities to receive title to, own, acquire, deliver, receive, possess, use, and transfer special nuclear material. This program is necessary to verify that companies can safely use special nuclear material before taking possession and starting operations. The inspection program's purpose is to obtain objective information that will permit the agency to assess whether its licensed fuel cycle facilities are operated safely and in compliance with regulations and that licensee activities do not pose undue safety and safeguards risks.

In early FY 2010, the NRC hired a management consulting firm to perform an independent evaluation of the agency's Fuel Facilities Licensing and Inspection (FFLI) Program. The scope of the study included issues relevant to how the FFLI Program contributes to the NRC's Safety and Security goals, including program purpose and design, strategic planning, program management, and program results and accountability. To develop the approach for the study, the contractor conducted an initial review of program activities and objectives. This initial review included a preliminary review of program documentation, authorizing legislation, and relevant regulations. The contractor also conducted a set of preliminary interviews with program staff and stakeholders. The purpose of this initial review was to identify key categories of program performance based on the

FFLI Program's legislative mandate, the NRC's rules and regulations, and the expectations and objectives important to the FFLI Program's stakeholders, constituents, and program staff.

The NRC received initial draft reports in June 2010, and the staff provided its comments on the initial drafts to the contractor in late August 2010. The contractor revised the initial drafts and provided a second set of drafts in early October 2010. Currently, the staff is reviewing these revisions, and will provide comments for the final report. The staff expects a final report in late November 2010.

Process Improvements

In order to make greater use of its resources and improve the efficiency, effectiveness, and timeliness of its processes, the NRC initiated a program that uses the Lean Six Sigma process improvement methodology. In FY 2010, the NRC successfully completed seven Lean Six Sigma process improvement trainings for qualified staff. Six NRC staff members received "Lean Six Sigma for Service" training at the level of "Black Belts." To increase the effectiveness of the NRC's process improvement initiatives, the NRC's Lean Six Sigma is also focused on streamlining the process by which the NRC conducts its process improvements along with collaborating, evaluating and assessing other agencies' best practices for possible use at the NRC.

Data Sources, Data Quality, and Data Security

The NRC's data collection and analysis methods are driven largely by the regulatory mandate that Congress entrusted to the agency. The NRC's mission is to regulate the Nation's civilian use of byproduct, source, and special nuclear materials to ensure adequate protection of public health and safety, protect the environment, and promote the common defense and security. In undertaking this mission, the agency oversees nuclear power plants, nonpower reactors, nuclear fuel facilities, interim spent fuel storage, radioactive material transportation, disposal of nuclear waste, and the industrial and medical uses of nuclear materials. Section 208 of the Energy Reorganization Act of 1974, as amended, requires the NRC to inform Congress of incidents or events that the Commission determines to be significant from the standpoint of public health and safety. The agency developed the Abnormal Occurrence Criteria to comply with the legislative intent of the Energy Reorganization Act to determine which events should be considered significant. Based on these criteria, the agency prepares an annual "Report to Congress on Abnormal Occurrences" (NUREG-0090; Volume 32 for FY 2009), issued June 2010, is available on the agency's public Web site at http://www.nrc.gov/reading-rm/doc-collections/nuregs/staff/sr0090/v32.

One important characteristic of this report is that the data presented normally originate from external sources, such as Agreement States and NRC licensees. The NRC finds these data credible because (1) agency regulations require Agreement States, licensees, and other external sources to report the necessary information, (2) the NRC maintains an aggressive inspection program that, among other activities, includes auditing licensee programs and evaluating Agreement State programs to ensure that they are reporting the necessary information as required by the agency's regulations, and (3) the NRC has established procedures for inspecting and evaluating licensees. The agency employs multiple database systems to support this process, including the Licensee Event Report Search System, the Accident Sequence Precursor Database, the Nuclear Materials Events Database, and the Radiation Exposure Information Report System. In addition, nonsensitive reports submitted by Agreement States and NRC licensees are available to the public through ADAMS, accessible through the agency's Web site (www.nrc.gov/reading-rm/adams.html).

The NRC verifies the reliability and technical accuracy of event information reported to the agency. The agency periodically inspects licensees and reviews Agreement State programs. In addition, NRC

Headquarters, the regional offices, and Agreement States hold periodic conference calls to discuss event information. The staff validates and verifies events identified as meeting the Abnormal Occurrence Criteria before reporting them to Congress.

Additionally, the NRC actively participates in Data.gov, a Federal Web site designed to increase public access to high-value, machine-readable datasets generated by the Executive Branch. The NRC published its first dataset in October 2009, and in response to the Open Government Directive, published three additional datasets in January 2010. The NRC will continue to encourage public feedback on its high-value information, and consistent with agency policy and guidance provided by Data.gov, will continue to add new datasets to its high-value dataset publication plan.

Performance Data Completeness and Reliability

In order to manage for results, it is essential that the NRC assess the completeness and reliability of its performance data. Comparisons of actual performance with the projected levels are possible only if the data used to measure performance are complete and reliable. Consequently, the Reports Consolidation Act of 2000 requires the NRC Chairman to assess the completeness and reliability of the performance data used in this report. The process for ensuring that the data are complete and reliable requires offices to complete a template for submission to the Chief Financial Officer for every performance measure certifying that the applicable office director has approved the data submitted.

Data Completeness

The NRC considers data to be complete if the agency reports actual performance data for every performance goal and indicator in the annual plan. Actual performance data include all data that are available when the agency sends its report to the President and Congress. The agency has reported actual data for every strategic and performance goal measure. As a result, the data presented in this report meet the requirements for data completeness.

Data Reliability

The NRC considers data to be reliable when agency managers and decisionmakers use the data in carrying out their responsibilities. The data presented in this report meet this requirement for data reliability because NRC managers and senior leaders regularly use the reported data in the course of their duties.

Information Security

The NRC's information security program (1) protects NRC and licensee information and information systems from unauthorized access, use, disclosure, disruption, modification, or destruction, (2) protects electronic control functions from unauthorized access or manipulation, and (3) ensures that adequate controls for protecting security-related information are used in the conduct of NRC business. The NRC information security program includes measures to accomplish the following:

(1) Ensure that information security requirements, standards, and guidance are clear, concise, appropriate, and able to mitigate the potential adverse effects if sensitive information is compromised.

(2) Ensure that security controls for information owned by or under the control of the NRC are consistent with established information security controls, that security controls for information are operating as intended, that they are having the desired impact, and that similar controls for licensees regulated by the NRC are in compliance with NRC information security regulations.

(3) Ensure that suspected or actual information security violations are evaluated and appropriate sanctions are considered.

(4) Ensure that the NRC has made sufficient preparations for information security-related emergencies and incidents.

(5) Ensure that internal information security program components complement each other and are periodically evaluated and improved.

Photo Courtesy of Elekta

Refueling operations being conducted at a commercial nuclear station.

Photo Courtesy of NRC Photo Library

Excavation for two proposed Westinghouse AP1000 reactors at the Southern Nuclear Operating Company's Vogtle facility in Waynesboro, GA - November 4, 2009.

Chapter 3

Financial Statements and Auditor's Report

Photo Courtesy of NRC Photo Library

Indian Point Energy Center, Buchanan, NY.

Photo Courtesy of NRC Photo Library

Chairman Gregory B. Jaczko at the groundbreaking ceremony for the new NRC Headquarters building on May 17, 2010. The 14-story, 362,000-square-foot building will provide office space for 1,300 to 1,400 NRC staff members, and will take approximately 27 months to complete.

A Message from the Chief Financial Officer

I am pleased to present the financial statements for the U.S. Nuclear Regulatory Commission (NRC) Fiscal Year (FY) 2010 Performance and Accountability Report. For the seventh consecutive year, an independent auditor has rendered an unqualified opinion on the NRC financial statements. The auditor also rendered an unqualified opinion on our internal controls concluding that NRC had no reportable conditions or significant deficiencies.

In FY 2010, the NRC completed the necessary development, testing and training to successfully transition to a new core financial system at the beginning of FY 2011. The NRC's new core financial system replaces five stand-alone financial systems with nine subsystems. In our continuing efforts to improve budget execution, NRC recovered over $20 million of unused funds from completed contracts during the past year. The agency also completed a major budget restructuring to better align funding with agency strategies. This new system and revised budget structure will play an integral role in making the NRC's financial management more transparent, efficient, and effective in the future.

In FY 2011, the NRC will continue to refine its processes to enhance its financial operations using the advancements implemented in recent years. We will also begin additional modifications to our core financial system to seamlessly align the agency's acquisition function with budget development and execution. The NRC also plans to modernize our Time and Attendance System to improve its usability. We will also update the agency's Strategic Plan to set clear high level direction and goals for the agency. The new Strategic Plan will provide an improved basis for determining the activities and resources needed in our performance budget.

The NRC is committed to ensuring the safety and security of the Nation's civilian use of nuclear materials in the most effective and efficient manner. The regulation of the Nation's expanding nuclear industry requires even more vigorous stewardship of limited taxpayer resources and demands superior financial performance. I am proud of the progress we have made in the past year to promote sound business practices in the conduct of our regulatory mission and am confident that the NRC will continue to make future improvements.

J.E. Dyer
Chief Financial Officer
November 12, 2010

Principal Statements

BALANCE SHEET
(In Thousands)

As of September 30,	2010	2009
Assets		
Intragovernmental		
Fund balance with Treasury (Note 2)	$ 420,080	$ 448,632
Accounts receivable (Note 3)	7,674	4,907
Other-Advances and prepayments	3,073	3,340
Total intragovernmental	430,827	456,879
Accounts receivable, net (Note 3)	123,242	123,217
Property and equipment, net (Note 4)	36,231	31,624
Other	25	32
Total Assets	$ 590,325	$ 611,752
Liabilities		
Intragovernmental		
Accounts payable	$ 13,876	$ 13,977
Other (Note 5)	5,986	5,489
Total intragovernmental	19,862	19,466
Accounts payable	26,666	37,023
Federal employee benefits (Note 6)	7,575	7,628
Other (Note 5)	106,041	80,639
Total Liabilities	160,144	144,756
Net Position		
Unexpended appropriations	311,869	338,637
Cumulative results of operations (Note 8)	118,312	128,359
Total Net Position	430,181	466,996
Total Liabilities and Net Position	$ 590,325	$ 611,752

The accompanying notes to the principal statements are an integral part of this statement.

STATEMENT OF NET COST
(In Thousands)

For the years ended September 30,	2010	2009
Nuclear Reactor Safety and Security		
Gross costs	$ 882,591	$ 796,898
Less: Earned revenue	(836,303)	(794,007)
Total Net Cost of Nuclear Reactor Safety and Security (Note 9)	46,288	2,891
Nuclear Materials and Waste Safety and Security		
Gross costs	257,862	245,961
Less: Earned revenue	(87,178)	(78,460)
Total Net Cost of Nuclear Materials and Waste Safety and Security (Note 9)	170,684	167,501
Net Cost of Operations	$ 216,972	$ 170,392

The accompanying notes to the principal statements are an integral part of this statement.

STATEMENT OF CHANGES IN NET POSITION
(In Thousands)

For the years ended September 30,	2010	2009
Cumulative Results of Operations		
Beginning Balance	$ 128,359	$ 128,235
Budgetary Financing Sources		
Appropriations used	137,113	89,309
Non-exchange revenue (Note 11)	-	-
Transfers-in/out without reimbursement	29,000	49,000
Other Financing Sources		
Imputed financing from costs absorbed by others (Note 11)	40,812	32,207
Total Financing Sources	206,925	170,516
Net Cost of Operations	(216,972)	(170,392)
Net Change	(10,047)	124
Cumulative Results of Operations	$ 118,312	$ 128,359
Unexpended Appropriations		
Beginning Balance	$ 338,637	$ 289,269
Budgetary Financing Sources		
Appropriations received	128,345	138,677
Other adjustments (Recissions)	(18,000)	-
Appropriations used	(137,113)	(89,309)
Total Budgetary Financing Sources	(26,768)	49,368
Total Unexpended Appropriations	311,869	338,637
Net Position	$ 430,181	$ 466,996

The accompanying notes to the principal statements are an integral part of this statement.

STATEMENT OF BUDGETARY RESOURCES
(In Thousands)

For the years ended September 30,	2010	2009
Budgetary Resources		
Unobligated balance, brought forward, October 1	$ 81,126	$ 78,990
Recoveries of prior year unpaid obligations		
Actual	22,446	28,371
Budget authority		
Appropriation	1,066,859	1,045,517
Spending authority from offsetting collections		
Reimbursements earned-collected	10,086	8,429
Reimbursements earned-change in receivables	(424)	375
Change in unfilled customer orders-advance received	1,198	333
Change in unfilled customer orders-without advance	493	3,190
Subtotal-spending authority from offsetting collections	11,353	12,327
Permanently not available	(18,000)	-
Total Budgetary Resources	$ 1,163,784	$ 1,165,205
Status of Budgetary Resources		
Obligations incurred (Note 12)		
Direct	$ 1,108,948	$ 1,073,782
Reimbursable	10,137	10,297
Subtotal	1,119,085	1,084,079
Unobligated balance		
Apportioned	29,744	66,699
Exempt from apportionment	7,079	7,609
Subtotal	36,823	74,308
Unobligated balance, not available	7,876	6,818
Total Status of Budgetary Resources	$ 1,163,784	$ 1,165,205
Change in Obligated Balance		
Obligated balance, net		
Unpaid obligations brought forward, October 1	$ 367,498	$ 314,488
Obligations incurred, net	1,119,085	1,084,079
Gross outlays	(1,088,687)	(999,133)
Recoveries of prior year unpaid obligations, actual	(22,446)	(28,371)
Change in uncollected customer payments, from Federal sources	(69)	(3,565)
Obligated balance, net, end of period		
Unpaid obligations	383,154	375,201
Uncollected customer payments, from Federal sources	(7,773)	(7,703)
Total unpaid obligated balance, net, end of period	$ 375,381	$ 367,498
Net outlays		
Gross outlays	$ 1,088,687	$ 999,133
Offsetting collections	(11,284)	(8,762)
Distributed offsetting receipts	(909,514)	(857,839)
Net Outlays	$ 167,889	$ 132,532

Notes to the Principal Statements
(All Tables are Presented in Thousands)

Note 1.
SUMMARY OF SIGNIFICANT ACCOUNTING POLICIES

A. Reporting Entity

The U.S. Nuclear Regulatory Commission (NRC) is an independent regulatory agency of the Federal Government that was created by the U.S. Congress to regulate the Nation's civilian use of byproduct, source, and special nuclear materials to ensure adequate protection of the public health and safety, to promote the common defense and security, and to protect the environment. Its purposes are defined by the Energy Reorganization Act of 1974, as amended, along with the Atomic Energy Act of 1954, as amended, which provide the foundation for regulating the Nation's civilian use of nuclear materials.

The NRC operates through the execution of its congressionally approved appropriations for Salaries and Expenses and the Office of the Inspector General, including funds derived from the Nuclear Waste Fund. In addition, the U.S. Agency for International Development (USAID) provides transfer appropriations to develop nuclear safety, regulatory authorities, and independent oversight of nuclear reactors in Russia, Ukraine, Kazakhstan, Georgia, and Armenia.

B. Basis of Presentation

These principal statements report the financial position and results of operations of the NRC as required by the Chief Financial Officers Act of 1990 and the Government Management Reform Act of 1994. These financial statements were prepared from the books and records of the NRC in conformance with generally accepted accounting principles (GAAP) of the United States and the form and content for entity financial statements specified by the Office of Management and Budget (OMB) in Circular

No. A-136, "Financial Reporting Requirements." GAAP for Federal entities are the standards prescribed by the Federal Accounting Standards Advisory Board, which is the official body for setting the accounting standards of the U.S. Government. These statements are, therefore, different from the financial reports, also prepared by the NRC pursuant to OMB directives, which are used to monitor and control the NRC's use of budgetary resources.

The NRC has not presented a Statement of Custodial Activity because the amounts involved are immaterial and incidental to its operations and mission.

C. Budgets and Budgetary Accounting

Budgetary accounting measures appropriation and consumption of budget spending authority or other budgetary resources and facilitates compliance with legal constraints and controls over the use of Federal funds. Under budgetary reporting principles, budgetary resources are consumed at the time of purchase. Assets and liabilities that do not consume current budgetary resources, are not reported, and only those liabilities for which valid obligations have been established are considered to consume budgetary resources.

For the past 36 years, Congress has enacted no-year appropriations, which are available for obligation by the NRC until expended. For FY 2010, the Energy and Water Development and Related Agencies Appropriations Act, 2010 requires the NRC to recover approximately 90 percent of its new budget authority by assessing fees for licensing and inspection activities.

D. Basis of Accounting

These financial statements reflect both accrual and budgetary accounting transactions. Under the accrual method, revenues are recognized when earned and expenses are recognized when a liability is incurred, without regard to receipt or payment of cash. Budgetary accounting is also used to record the obligation of funds prior to the accrual-based transaction. The Statement of Budgetary Resources presents budgetary resources available to the NRC and changes in obligations during the year. Interest on borrowings of the U.S. Department of the Treasury (Treasury) is not included as a cost to NRC programs and is not included in the accompanying financial statements.

E. Revenues and Other Financing Sources

The NRC is required to offset its appropriations by revenue received during the fiscal year from the assessment of fees. The NRC assesses two types of fees to recover its budget authority: (1) fees assessed under Title 10 of the *Code of Federal Regulations* (10 CFR) Part 170, "Fees for Facilities, Materials, Import and Export Licenses, and Other Regulatory Services under the Atomic Energy Act of 1954, as Amended," for licensing, inspection, and other services under the authority of the Independent Offices Appropriation Act of 1952 to recover the NRC's costs of providing individually identifiable services to specific applicants and licensees and (2) annual fees assessed for nuclear facilities and materials licensees under 10 CFR Part 171, "Annual Fees for Reactor Licenses and Fuel Cycle Licenses and Material Licenses." Licensing revenues are recognized on a straight-line basis over the licensing period. Inspection fees are recorded as revenues when the services are performed.

For accounting purposes, appropriations are recognized as financing sources (appropriations used) at the time goods and services are received. At the end of the fiscal year, appropriations recognized are reduced by the amount of assessed fees collected during the fiscal year to the extent of new budget authority for the year. Collections that exceed the new budget authority are held to offset subsequent years' appropriations. Appropriations expended for property and equipment are recognized as expenses when the asset is consumed in operations as reflected by depreciation and amortization expense.

F. Fund Balance with Treasury

The NRC's cash receipts and disbursements are processed by the Treasury. The Fund Balance with Treasury is primarily appropriated funds that are available to pay current liabilities and to finance authorized purchase commitments. The Fund Balance with Treasury represents the NRC's right to draw on the Treasury for allowable expenditures.

G. Accounts Receivable

Accounts receivable consist of amounts owed to the NRC by other Federal agencies and the public. Amounts due from the public are presented net of an allowance for uncollectible accounts. The allowance is determined based on the age of the receivable and allowance rates established from historical experience. Receivables from Federal agencies are expected to be collected; therefore, there is no allowance for uncollectible accounts for Federal agencies.

H. Non-Entity Assets

Non-entity assets consist of miscellaneous penalties and interest due from the public, which, when collected, must be transferred to the Treasury.

I. Property and Equipment

Property and equipment consist primarily of typical office furnishings, leasehold improvements, nuclear reactor simulators, and computer hardware and software. The costs of internal use software include the full cost of salaries and benefits for agency personnel involved in software development. The NRC has no real property. The land and buildings in which the NRC operates are provided by the General Services Administration (GSA), which charges the NRC rent that approximates the commercial rental rates for similar properties.

Property with a cost of $50 thousand or more per unit and a useful life of 2 years or more is capitalized at cost and depreciated using the straight-line method over the useful life. Other property items are expensed when purchased. Normal repairs and maintenance are charged to expense as incurred.

J. Accounts Payable

The NRC uses an estimation methodology to calculate the accounts payable balance which represents costs for billed and unbilled goods and services received (prior to year end) that are unpaid. The NRC uses available information from program staff for a majority of the NRC's largest obligations and uses an algorithm to estimate the liability for smaller obligation balances. This estimation methodology is validated quarterly.

K. Liabilities Not Covered by Budgetary Resources

Liabilities represent the amount of monies or other resources that are likely to be paid by the NRC as the result of a transaction or event that has already occurred. No liability can be paid by the NRC absent an appropriation. Liabilities for which an appropriation has not been enacted are classified as "Liabilities Not Covered by Budgetary Resources." Also, the NRC liabilities arising from sources other than contracts can be abrogated by the Government acting in its sovereign capacity.

Intragovernmental

The NRC records a liability to the U.S. Department of Labor (DOL) for Federal Employees Compensation Act (FECA) benefits paid by DOL on behalf of the NRC.

Federal Employee Benefits

Federal employee benefits represent the actuarial liability for estimated future FECA disability benefits. The future workers' compensation estimate was generated by DOL from an application of actuarial procedures developed to estimate the liability for FECA, which includes the expected liability for death, disability, medical, and miscellaneous costs for approved compensation cases. The liability is calculated using historical benefit payment patterns related to a specific incurred period to predict the ultimate payments related to that period. These projected annual benefit payments are discounted to present value. The interest rate assumptions utilized for discounting benefits are 3.65 percent and 4.22 percent for FY 2010 and FY 2009, respectively.

Other

Accrued annual leave represents the amount of annual leave earned by NRC employees but not yet taken.

L. Contingencies

Contingent liabilities are those for which the existence or amount of the liability cannot be determined with certainty pending the outcome of future events. The uncertainty should ultimately be resolved when one or more future events occur or fail to occur. A contingent liability (included in Other Liabilities) should be recorded when a past event or exchange transaction has occurred; a future outflow or other sacrifice of resources is probable; and the future outflow or sacrifice of resources is measurable. A contingency is considered probable when the future confirming event or events are more likely than not to occur, with the exception of pending or threatened litigation and unasserted claims. A contingency is disclosed in the Notes to the Financial Statements if any of the conditions for liability recognition are not met and there is at least a reasonable possibility that a loss or an additional loss may have been incurred (Note 16). A contingency is considered reasonably possible when the chance of the future confirming event or events occurring is more than remote but less than probable. A contingency is not recognized as a contingent liability and an expense nor disclosed in the Notes to the Financial Statements when the chance of the future event or events occurring is remote. A contingency is considered remote when the chance of the future event or events occurring is slight.

M. Annual, Sick, and Other Leave

Annual leave is accrued as it is earned and the accrual is reduced as leave is taken. Each year, the balance in the accrued annual leave liability account is adjusted to reflect current pay rates. To the extent that current or prior year funding is not available to cover annual leave earned but not taken, funding will be obtained from future financing sources. Sick leave and other types of nonvested leave are expensed as taken.

N. Retirement Plans

The NRC employees belong to either the Federal Employees Retirement System (FERS) or the Civil Service Retirement System (CSRS). For FY 2010 and FY 2009, for employees belonging to FERS, the NRC withheld 0.8 percent of base pay earnings, in addition to Federal Insurance Contribution Act (FICA) withholdings, and matched the withholdings with an 11.2 percent contribution. The sum is transferred to the Federal Employees Retirement Fund. For employees covered by CSRS, the NRC withholds 7 percent of base pay earnings. The NRC matched this withholding with a 7 percent contribution in FY 2010 and FY 2009.

The Thrift Savings Plan (TSP) is a retirement savings and investment plan for employees belonging to either FERS or CSRS. The maximum percentage of base pay that an employee participating in FERS or CSRS may contribute is unlimited in 2010 and 2009, subject to the maximum contribution of $16.5 thousand in 2010 and $16.5 thousand in 2009. For employees participating in FERS, the NRC automatically contributes 1 percent of base pay to their account and matches contributions up to an additional 4 percent. For employees participating in CSRS, there is no NRC matching of the contribution. The sum of the employees' and the NRC's contributions is transferred to the Federal Retirement Thrift Investment Board.

The NRC does not report on its financial statements FERS and CSRS assets, accumulated plan benefits, or unfunded liabilities, if any, applicable to its employees. Reporting such amounts is the responsibility of the U.S. Office of Personnel Management. The portion of the current and estimated future outlays for CSRS not paid by the NRC is included in the NRC's financial statements as an imputed financing source in the NRC's Statement of Changes in Net Position and as program costs on the Statement of Net Cost.

O. Leases

The NRC's capital leases are for personal property consisting of reproduction equipment which is installed at NRC Headquarters. For FY 2010, there are eight capital leases with terms of 5 years, consisting of two capital leases added in FY 2008 with an interest rate of 3.99 percent, two capital leases that were added in FY 2007 with an interest rate of 4.58 percent, one capital lease in FY 2006 with an interest rate of 4.25 percent, and three capital leases for FY 2005 with an interest rate of 4.13 percent. The reproduction equipment is depreciated over 5 years using the straight-line method with no salvage value.

Operating leases consist of real property leases with GSA. The leases are for NRC's headquarters and regional offices. The GSA charges the NRC lease rates which approximate commercial rates for comparable space.

P. Pricing Policy

The NRC provides nuclear reactor and materials licensing and inspection services to the public and other Government entities. In accordance with OMB Circular No. A-25, "User Charges," and the Independent Offices Appropriation Act of 1952, the NRC assesses fees under 10 CFR Part 170 for licensing and inspection activities to recover the full cost of providing individually identifiable services.

The NRC's policy is to recover the full cost of goods and services provided to other Government entities where (1) the services performed are not part of its statutory mission and (2) the NRC has not received appropriations for those services. Fees for reimbursable work are assessed at the 10 CFR Part 170 rate with minor exceptions for programs that are nominal activities of the NRC.

Q. Net Position

The NRC's net position consists of unexpended appropriations and cumulative results of operations. Unexpended appropriations represent appropriated spending authority that is unobligated and has not been withdrawn by the Treasury and obligations that have not been paid. Cumulative results of operations represent the excess of financing sources over expenses since inception.

R. Use of Management Estimates

The preparation of the accompanying financial statements in accordance with generally accepted accounting principles requires management to make certain estimates and assumptions that affect the reported amounts of assets, liabilities, revenues, and expenses. Actual results could differ from those estimates.

S. Appropriation Transfers

The NRC is a party to allocation transfers with the U.S. Agency for International Development (USAID) as a receiving (child) entity. These transfers are for the international development of nuclear safety and regulatory authorities in Russia, Ukraine, Kazakhstan, Georgia, and Armenia for the startup, operation, shutdown, and decommissioning of Soviet-designed nuclear power plants; the safe and secure use of radioactive materials; and the accounting for and protection of nuclear materials. Allocation transfers are legal delegations by one agency of its authority to obligate budget authority and outlay funds to another agency. All financial activity related to these allocation transfers (e.g., budget authority, obligations, outlays) is reported in the financial statements of the parent entity from which the underlying legislative authority, appropriations, and budget apportionments are derived. The NRC receives allocation transfers, as the child, from USAID.

T. Statement of Net Cost

The programs as presented on the Statement of Net Cost are based on the annual performance budget and are described as follows:

The Nuclear Reactor Safety and Security program encompasses all NRC efforts to ensure that civilian nuclear power reactor facilities and research and test reactors are licensed and operated in a manner that adequately protects the public health and safety, and the environment, and protects against radiological sabotage and theft or diversion of special nuclear materials. The Nuclear Reactor Safety and Security program contains the following activities: new reactors, reactor licensing tasks, reactor license renewal, international activities, reactor oversight, and incident response.

The Nuclear Materials and Waste Safety and Security encompasses all NRC efforts to protect the public health and safety and the environment and ensures the secure use and management of radioactive materials. The Nuclear Materials and Waste Safety and Security program contains the following activities: fuel facilities, nuclear materials users, decommissioning and low-level waste, spent fuel storage and transportation, and high-level waste repository.

For intragovernmental gross costs, the buyers and sellers are both Federal entities. For earned revenues from the public, the buyers of the goods or services are non-Federal entities.

Note 2. FUND BALANCE WITH TREASURY

	2010	2009
Fund Balances		
Appropriated funds	$ 400,435	$ 423,724
Nuclear Waste Fund	19,645	24,900
Other fund types	-	8
Total	$ 420,080	$ 448,632
Status of Fund Balance with Treasury		
Unobligated balance		
Available		
Appropriated funds	$ 36,823	$ 74,308
Unavailable	7,876	6,818
Obligated balance not yet disbursed	375,381	367,498
Non-budgetary funds with Treasury	-	8
Total	$ 420,080	$ 448,632

The Fund Balance with Treasury consists of unobligated and obligated balance budgetary accounts. It includes Nuclear Waste Fund activity. The Nuclear Waste Fund unobligated balance is $7.1 million and $7.6 million as of September 30, 2010, and 2009, respectively.

Note 3. ACCOUNTS RECEIVABLE

	2010	2009
Intragovernmental		
Fee receivables and reimbursements	$ 7,674	$ 4,907
Receivables with the Public		
Materials and facilities fees-billed	$ 2,611	$ 3,316
Materials and facilities fees-unbilled	123,416	122,929
Other	77	113
Total Receivables with the Public	126,104	126,358
Less: Allowance for uncollectible accounts	(2,862)	(3,141)
Total Receivables with the Public, Net	$ 123,242	$ 123,217
Total Accounts Receivable	$ 133,778	$ 131,265
Less: Allowance for uncollectible accounts	(2,862)	(3,141)
Total Accounts Receivable, Net	$ 130,916	$ 128,124

Note 4. PROPERTY AND EQUIPMENT, NET

Fixed Assets Class	Service Years	Acquisition Value	Accumulated Depreciation and Amortization	2010 Net Book Value	2009 Net Book Value
Equipment	5-8	$ 13,188	$ (11,247)	$ 1,941	$ 1,365
Leased equipment	5-8	1,712	(1,154)	558	896
IT software	5	53,866	(45,799)	8,067	11,956
IT software under development	-	5,153	-	5,153	2,227
Leasehold improvements	20	38,250	(24,210)	14,040	14,727
Leasehold improvements in progress	-	6,472	-	6,472	453
Total		$ 118,641	$ (82,410)	$ 36,231	$ 31,624

Note 5. OTHER LIABILITIES

	2010	2009
Intragovernmental		
Liability to offset miscellaneous accounts receivable	$ 6	$ 40
Liability for advances from other agencies	82	88
Accrued workers' compensation	1,719	1,725
Accrued unemployment compensation	31	25
Employee benefit contributions	4,148	3,611
Total Intragovernmental Other Liabilities	$ 5,986	$ 5,489
Other Liabilities		
Accrued annual leave	$ 50,413	$ 47,271
Accrued salaries and benefits	26,621	23,134
Contract holdbacks, advances, capital lease liability, and other	7,391	7,155
Contingent liabilities	11,750	-
Grants payable	9,866	3,079
Total Other Liabilities	$ 106,041	$ 80,639
Total Intragovernmental and Other Liabilities	$ 112,027	$ 86,128

Other liabilities are current except for capital lease liability (Note 7).

Note 6. LIABILITIES NOT COVERED BY BUDGETARY RESOURCES

	2010	2009
Intragovernmental		
FECA paid by DOL	$ 1,719	$ 1,725
Accrued unemployment compensation	31	25
Federal Employee Benefits		
Future FECA	7,575	7,628
Other		
Accrued annual leave	50,413	47,271
Contingent liabilities	11,750	-
Total Liabilities not Covered by Budgetary Resources	71,488	56,649
Total Liabilities Covered by Budgetary Resources	88,656	88,107
Total Liabilities	$ 160,144	$ 144,756

Liabilities not Covered by Budgetary Resources represents the amount of future funding needed to pay the accrued unfunded expenses as of September 30, 2010, and 2009. These liabilities are not funded from current or prior-year appropriations and assessments, but rather should be funded from future appropriations and assessments. Accordingly, future funding requirements have been recognized for the expenses that will be paid from future appropriations.

Note 7. LEASES

		2010	2009
Assets under capital leases:			
Copiers and booklet maker		$ 1,712	$ 1,712
Accumulated depreciation		(1,154)	(816)
Net assets under capital leases		$ 558	$ 896

			2010	2009
Future Lease Payments Due: Fiscal Year	**Capital**	**Operating**		
2010	$ -	$ -	$ -	$ 32,882
2011	308	31,339	31,647	32,637
2012	272	29,580	29,852	30,508
2013	14	24,740	24,754	22,624
2014	-	10,546	10,546	8,550
2015 and thereafter	-	39,198	39,198	17,443
Total Lease Liability	594	135,403	135,997	144,644
Add: Imputed Interest	27	-	27	60
Total Future Lease Payments	$ 621	$ 135,403	$ 136,024	$ 144,704

The Capital Lease Liability of $594 thousand is included in Other Liabilities (Note 5).

NOTE 8. CUMULATIVE RESULTS OF OPERATIONS

	2010	2009
Liabilities not covered by budgetary resources (Note 6)	$ (71,488)	$ (56,649)
Investment in property and equipment, net (Note 4)	36,231	31,624
Contributions from foreign cooperative research agreements	3,632	2,606
Nuclear Waste Fund	19,592	23,703
Accounts receivable - fees	130,300	127,020
Other	45	55
Cumulative Results of Operations	**$ 118,312**	**$ 128,359**

NOTE 9. STATEMENT OF NET COST

For the years ended September 30,	2010	2009
Nuclear Reactor Safety and Security		
Intragovernmental gross costs	$ 272,871	$ 238,234
Less: Intragovernmental earned revenue	(54,270)	(39,307)
Intragovernmental net costs	218,601	198,927
Gross costs with the public	609,720	558,664
Less: Earned revenues from the public	(782,033)	(754,700)
Net costs with the public	(172,313)	(196,036)
Total Net Cost of Nuclear Reactor Safety and Security	**$ 46,288**	**$ 2,891**
Nuclear Materials and Waste Safety and Security		
Intragovernmental gross costs	$ 64,260	$ 59,253
Less: Intragovernmental earned revenue	(7,314)	(6,190)
Intragovernmental net costs	56,946	53,063
Gross costs with the public	193,602	186,708
Less: Earned revenues from the public	(79,864)	(72,270)
Net costs with the public	113,738	114,438
Total Net Cost of Nuclear Materials and Waste Safety and Security	**$ 170,684**	**$ 167,501**

NOTE 10. EXCHANGE REVENUES

	2010	2009
Fees for licensing, inspection, and other services	$ 912,794	$ 864,155
Revenue from reimbursable work	10,687	8,312
Total Exchange Revenues	**$ 923,481**	$ 872,467

Note 11. FINANCING SOURCES OTHER THAN EXCHANGE REVENUE

	2010	2009
Appropriations Used		
Collections were used to reduce the fiscal year's appropriations recognized:		
Funds consumed	$ 1,079,739	$ 993,884
Less: Collection from fees assessed	(909,514)	(857,839)
Less: Nuclear Waste Funding expense	(33,112)	(46,736)
Total Appropriations Used	**$ 137,113**	$ 89,309

Funds consumed include $81.1 million and $78.9 million through
September 30, 2010, and 2009 respectively, of available funds from prior years.

	2010	2009
Non-Exchange Revenue		
Civil penalties	$ 590	$ 278
Miscellaneous receipts	879	108
Contra-Revenue	(1,469)	(386)
Total Non-Exchange Revenue	**$ -**	$ -

	2010	2009
Imputed Financing		
Civil Service Retirement System	$ 19,895	$ 11,258
Federal Employee Health Benefit	20,825	19,898
Federal Employee Group Life Insurance	92	88
Judgments/Awards	-	963
Total Imputed Financing	**$ 40,812**	$ 32,207

Note 12. TOTAL OBLIGATIONS INCURRED

	2010	2009
Direct Obligations		
Category A	$ 1,079,158	$ 1,022,122
Exempt from Apportionment	29,790	51,660
Total Direct Obligations	1,108,948	1,073,782
Reimbursable Obligations	10,137	10,297
Total Obligations Incurred	$ 1,119,085	$ 1,084,079

Obligations exempt from apportionment are the result of funds derived from the Nuclear Waste Fund. Category A Obligations consist of NRC appropriations only. Undelivered orders for the Nuclear Waste Fund are $12.5 million and $16.1 million, Salaries and Expenses $288.1 million and $276.2 million, and the Office of the Inspector General $1.2 million and $2.3 million through September 30, 2010, and 2009, respectively.

Note 13. NUCLEAR WASTE FUND

Included in NRC's budget for FY 2010 and 2009 are $29 million and $49 million, respectively, provided from the Nuclear Waste Fund. Statement of Federal Financial Accounting Standards (SFFAS) No. 27, "Identifying and Reporting Earmarked Funds," lists three defining criteria for an earmarked fund. Generally, an earmarked fund is established by law to use specifically identified financing sources only for designated activities, and the statute provides explicit authority to retain current, unused revenues for future use. Also, the law includes a requirement to account for and report on the receipt and use of the financing sources as distinguished from general revenues.

In 1982, Congress passed the Nuclear Waste Policy Act of 1982 (Public Law 97-425) establishing the Nuclear Waste Fund (NWF) to be administered by the U.S. Department of Energy (DOE) (42 U.S.C. 10222). Given the terms of the statute, the NWF clearly meets the definition of an earmarked fund from DOE's perspective, and DOE does indeed report the NWF as an earmarked fund in its Performance and Accountability Report (PAR).

For the NRC, the NWF transfer is a source of financing; its receipt of NWF funds is a use of NWF resources. The NRC collects no revenue on behalf of the NWF and has no administrative control over it. Furthermore, the Treasury has no separate fund symbol for the NWF under the NRC's agency location code (ALC). The receipt and expenditure of NWF money are reported to Treasury under the NRC's primary Salaries and Expenses fund (X0200).

Based on these facts, the NWF is not an earmarked fund from the NRC's perspective. To provide additional information to the users of these financial statements, enhanced disclosure of the fund is presented below.

The funding provided to the NRC in FY 2010 and FY 2009 was for the purpose of performing activities associated with DOE's application for a high-level waste repository at Yucca Mountain, NV. These activities included assistance to DOE with the application, review of the application, conduct of thorough safety and security evaluations, preparation of the safety evaluation report, initiation of the inspection program, ensuring that the regulation process was made available to stakeholders and the general public, and providing legal advice and representation for staff reviews and Commission actions.

The NWF amounts received, expended, obligated, and unobligated balances as of September 30, 2010, and 2009, are shown in the following:

	2010	2009
Appropriations received	$ 29,000	$ 49,000
Expended appropriations	$ 34,308	$ 47,062
Obligations incurred	$ 29,790	$ 51,660
Unobligated balances	$ 7,079	$ 7,608

Note 14. EXPLANATION OF DIFFERENCES BETWEEN THE STATEMENT OF BUDGETARY RESOURCES AND THE BUDGET OF THE U.S. GOVERNMENT

Statement of Federal Financial Standards (SFFAS) No. 7, "Accounting for Revenue and Other Financing Sources," requires the NRC to reconcile the budgetary resources reported on the Statement of Budgetary Resources to the prior fiscal year actual budgetary resources presented in the Budget of the U.S. Government and explain any material differences. The NRC does not have any material differences between the Statement of Budgetary Resources and the Budget of the U.S. Government. The President's Budget with actual results for the NRC has not been published for FY 2010. It is expected to be published February 2011.

Note 15. RECONCILIATION OF NET COST OF OPERATIONS TO BUDGETARY RESOURCES

For the years ended September 30,	2010	2009
Budgetary Resources Obligated		
Obligations incurred (Note 12)	$ 1,119,085	$ 1,084,079
Less: Spending authority from offsetting collections and recoveries	(33,799)	(40,698)
Less: Distributed offsetting receipts	(909,514)	(857,839)
Net Obligations	175,772	185,542
Other Resources		
Imputed financing from costs absorbed by others	40,812	32,207
Net Other Resources Used to Finance Activities	40,812	32,207
Total Resources Used to Finance Activities	216,584	217,749
Resources Used to Finance Items not Part of the Net Cost of Operations	(19,668)	(53,413)
Total Resources Used to Finance the Net Cost of Operations	196,916	164,336
Components of the Net Cost of Operations that will not require or generate resources in the current period	20,056	6,056
Net Cost of Operations	$ 216,972	$ 170,392

Note 16. CONTINGENCIES

The NRC is subject to potential liabilities in various administrative proceedings, legal actions, environmental suits, and claims brought against it. In the opinion of the NRC's management and legal counsel, the ultimate resolution of these proceedings, actions, suits, and claims will not materially affect the financial position or net costs of the NRC.

Probable Likelihood of an Adverse Outcome:

The NRC is subject to potential liabilities where adverse outcomes are probable, and claims are approximately $11.8 million as of September 30, 2010. Accordingly, $11.8 million of contingent liabilities were included in Other Liabilities on the Consolidated Balance Sheet as of September 30, 2010. Any amounts ultimately due for these claims will be paid out of Treasury's Judgment Fund. Once the claims are settled or court judgments are assessed, the liability will be removed and an Imputed Financing Source from Costs Absorbed by Others will be recognized.

Reasonably Possible Likelihood of an Adverse Outcome:

The NRC is subject to potential liabilities where adverse outcomes are reasonably possible. The upper range of loss on these potential liabilities is $150 thousand.

Required Supplementary Information
Schedule of Budgetary Resources (In Thousands)

For the fiscal year ended September 30, 2010	Salaries and Expenses	Office of Inspector General	Nuclear Facility Fees	Total
	X0200	X0300	X5280	
Budgetary Resources				
Unobligated balances, brought forward, October 1	$ 79,657	$ 1,468	$ 1	$ 81,126
Recoveries of prior year obligations				
Actual	21,199	1,247	-	22,446
Budget authority				
Appropriation	1,056,000	10,860	(1)	1,066,859
Spending authority from offsetting collections				
Reimbursements earned-collected	10,086	-	-	10,086
Reimbursements earned-change in receivables	(424)	-	-	(424)
Change in unfilled customer orders-advance received	1,198	-	-	1,198
Change in unfilled customer orders-without advance	493	-	-	493
Subtotal-spending authority from offsetting collections	11,353	-	-	11,353
Permanently not available	(18,000)	-	-	(18,000)
Total Budgetary Resources	**$ 1,150,209**	**$ 13,575**	**$ -**	**$ 1,163,784**
Status of Budgetary Resources				
Obligations incurred (Note 12)				
Direct	$ 1,097,260	$ 11,688	$ -	$ 1,108,948
Reimbursable	10,137	-	-	10,137
Subtotal	1,107,397	11,688	-	1,119,085
Unobligated balance				
Apportioned	28,654	1,090	-	29,744
Exempt from apportionment	7,079	-	-	7,079
Subtotal	35,733	1,090	-	36,823
Unobligated balance, not available	7,079	797	-	7,876
Total Status of Budgetary Resources	**$ 1,150,209**	**$ 13,575**	**$ -**	**$ 1,163,784**
Change in Obligated Balance				
Obligated balance, net				
Unpaid obligations, brought forward, October 1	$ 365,851	$ 1,647	$ -	$ 367,498
Obligations incurred, net	1,107,397	11,688	-	1,119,085
Gross outlays	(1,077,555)	(11,132)	-	(1,088,687)
Recoveries of prior year obligations, actual	(21,199)	(1,247)	-	(22,446)
Change in uncollected customer payments, from Federal sources	(69)	-	-	(69)
Obligated balance, net, end of period				
Unpaid obligations	382,198	956	-	383,154
Uncollected customer payments, from Federal sources	(7,773)	-	-	(7,773)
Total unpaid obligated balance, net, end of period	$ 374,425	$ 956	$ -	$ 375,381
Net outlays				
Gross outlays	$ 1,077,555	$ 11,132	$ -	$ 1,088,687
Offsetting collections	(11,284)	-	-	(11,284)
Distributed offsetting receipts	-	-	(909,514)	(909,514)
Net Outlays	**$ 1,066,271**	**$ 11,132**	**$ (909,514)**	**$ 167,889**

Photo Courtesy of NRC Photo Library

Progress Energy staff discuss the status the Crystal River Unit 3 containment wall with Deputy Executive Director for Reactor and Preparedness Programs Martin J. Virgillio, Region II Division Director of Reactor Projects Len Wert, and Sr. Resident Inspector Tom Morrissey - July 2010.

Auditor's Report

UNITED STATES
NUCLEAR REGULATORY COMMISSION
WASHINGTON, D.C. 20555-0001

OFFICE OF THE
INSPECTOR GENERAL

November 9, 2010

MEMORANDUM TO: Chairman Jaczko

FROM: Hubert T. Bell **/RA/**
Inspector General

SUBJECT: RESULTS OF THE AUDIT OF THE UNITED STATES
NUCLEAR REGULATORY COMMISSION'S FINANCIAL
STATEMENTS FOR FISCAL YEARS 2010 and 2009
(OIG-11-A-04)

The Chief Financial Officers Act of 1990, as amended (CFO Act), requires the Inspector
General (IG) or an independent external auditor, as determined by the IG, to annually
audit the United States Nuclear Regulatory Commission's (NRC) financial statements in
accordance with applicable standards. In compliance with this requirement, Urbach
Kahn & Werlin, LLP (UKW) was retained by the Office of the Inspector General (OIG) to
conduct this annual audit. Transmitted with this memorandum are the following UKW
reports:

- Opinion on the Principal Statements.

- Opinion on Internal Control.

- Compliance with Laws and Regulations.

NRC's Performance and Accountability Report includes comparative financial
statements for FY 2010 and FY 2009.

Objective of a Financial Statement Audit

The objective of a financial statement audit is to determine whether the audited entity's
financial statements are free of material misstatement. An audit includes examining, on
a test basis, evidence supporting the amounts and disclosures in the financial
statements. An audit also includes assessing the accounting principles used and
significant estimates made by management as well as evaluating the overall financial
statement presentation.

2

UKW's audit and examination were made in accordance with auditing standards generally accepted in the United States of America; *Government Auditing Standards* issued by the Comptroller General of the United States; attestation standards established by the American Institute of Certified Public Accountants; and Office of Management and Budget (OMB) Bulletin No. 07-04, *Audit Requirements for Federal Financial Statements*, as amended. The audit included, among other things, obtaining an understanding of NRC and its operations, including internal control over financial reporting; evaluating the design and operating effectiveness of internal control and assessing risk; and testing relevant internal controls over financial reporting. Because of inherent limitations in any internal control, misstatements due to error or fraud may occur and not be detected. Also, projections of any evaluation of the internal control to future periods are subject to the risk that the internal control may become inadequate because of changes in conditions, or that the degree of compliance with the policies or procedures may deteriorate.

FY 2010 Audit Results

The results are as follows:

> <u>Financial Statements</u>
>
> * Unqualified opinion
>
> <u>Internal Controls</u>
>
> * Unqualified opinion
>
> <u>Compliance with Laws and Regulations</u>
>
> * No reportable instances of noncompliance/no substantial noncompliance noted

Office of the Inspector General Oversight of UKW Performance

To fulfill our responsibilities under the CFO Act and related legislation for ensuring the quality of the audit work performed, we monitored UKW's audit of NRC's FY 2010 and FY 2009 financial statements by:

* Reviewing UKW's audit approach and planning.

* Evaluating the qualifications and independence of UKW's auditors.

* Monitoring audit progress at key points.

* Examining the working papers related to planning and performing the audit and assessing NRC's internal controls.

* Reviewing UKW's audit reports to ensure compliance with *Government Auditing Standards* and OMB Bulletin No. 07-04, as amended.

3

- Coordinating the issuance of the audit reports.

- Performing other procedures deemed necessary.

UKW is responsible for the attached auditor's reports, dated November 7, 2010, and the conclusions expressed therein. OIG is responsible for technical and administrative oversight regarding the firm's performance under the terms of the contract. Our review, as differentiated from an audit in conformance with *Government Auditing Standards*, was not intended to enable us to express, and accordingly we do not express, an opinion on:

- NRC's financial statements.

- The effectiveness of NRC's internal control over financial reporting.

- NRC's compliance with laws and regulations.

However, our monitoring review, as described above, disclosed no instances where UKW did not comply, in all material respects, with applicable auditing standards.

Meeting with the Chief Financial Officer

At the exit conference on November 8, 2010, representatives of the Office of the Chief Financial Officer, OIG, and UKW discussed the results of the audit.

Comments of the Chief Financial Officer

In his response, the Chief Financial Officer (CFO) agreed with the report. The full text of the CFO's response follows this report.

We appreciate NRC staff's cooperation and continued interest in improving financial management within NRC.

Attachment: As stated

cc: Commissioner Svinicki
        Commissioner Apostolakis
        Commissioner Magwood
        Commissioner Ostendorff
        N. Mamish, OEDO
        M. Muessle, OEDO
        J. Andersen, OEDO
        C. Jaegers, OEDO

U.S.NRC
UNITED STATES NUCLEAR REGULATORY COMMISSION

Independent Auditor's Report

UK&W Urbach Kahn & Werlin LLP
CERTIFIED PUBLIC ACCOUNTANTS

Inspector General
United States Nuclear Regulatory Commission

Chairman
United States Nuclear Regulatory Commission

We have audited the accompanying balance sheets of the United States Nuclear Regulatory Commission (NRC), as of September 30, 2010 and 2009, and the related statements of net cost, changes in net position, and budgetary resources (Principal Statements) for the years then ended. We also examined the NRC's internal control over financial reporting as of September 30, 2010 and 2009.

Summary

We concluded that the NRC's fiscal year (FY) 2010 Principal Statements are presented fairly, in all material respects, in conformity with accounting principles generally accepted in the United States of America. We also concluded that the NRC maintained, in all material respects, effective internal control over financial reporting. We noted no reportable instances of noncompliance with laws and regulations and no substantial noncompliance with federal financial management systems requirements, applicable Federal accounting standards, and the United States Government Standard General Ledger (USSGL) at the transaction level.

The following sections discuss in more detail: (1) these conclusions and our conclusions relating to other information presented in the Performance and Accountability Report, (2) management's responsibilities, and (3) our objectives, scope and methodology.

Opinion on the Principal Statements

In our opinion, the Principal Statements referred to above present fairly, in all material respects, the financial position of the NRC as of September 30, 2010 and 2009, and its net cost, changes in net position, and budgetary resources for the years then ended, in conformity with accounting principles generally accepted in the United States of America.

Opinion on Internal Control

In our opinion, the NRC maintained, in all material respects, effective control over financial reporting as of September 30, 2010, that provided reasonable assurance that misstatements, losses or noncompliance material in relation to the financial statements would be prevented, or detected and corrected, on a timely basis. Our opinion is based on criteria established under 31 U.S.C. 3512 (c), (d), the Federal Managers' Financial Integrity Act (FMFIA).

Compliance with Laws and Regulations

The results of our tests of compliance with laws and regulations disclosed no instances of noncompliance that are required to be reported under Government Auditing Standards and OMB Bulletin No. 07-04, Audit Requirements for Federal Financial Statements, as amended. Providing an opinion on compliance with laws and regulations was not an objective of our audit and, accordingly, we do not express such an opinion.

Under the Federal Financial Management Improvement Act (FFMIA), we are required to report whether the NRC's financial management systems substantially comply with the federal financial management systems requirements, applicable Federal accounting standards, and the United States Government Standard General Ledger (USSGL) at the transaction level. To meet this requirement, we performed tests of compliance with the provisions of FFMIA section 803(a). The results of our tests disclosed no substantial noncompliance with federal financial management systems requirements, applicable Federal accounting standards, and the USSGL at the transaction level.

Other Information

The information in Management's Discussion and Analysis and other Required Supplementary Information (RSI) in NRC's Performance and Accountability Report is not a required part of the Principal Statements, but is supplementary information required by accounting principles generally accepted in the United States of America. We have applied certain limited procedures, which consisted principally of inquiries of management regarding the methods of measurement and presentation of the supplementary information. However, we did not audit the information and express no opinion on it.

The Program Performance and Other Accompanying Information sections listed in the Table of Contents are presented for additional analysis and are not a required part of the Principal Statements. Such information has not been subjected to the auditing procedures applied in the audit of the financial statements and, accordingly, we express no opinion on them.

U.S.NRC

UNITED STATES NUCLEAR REGULATORY COMMISSION

Management Responsibilities

Management is responsible for (1) preparing the Principal Statements in conformity with accounting principles generally accepted in the United States of America, (2) establishing and maintaining effective internal control over financial reporting, and evaluating its effectiveness, (3) ensuring that the NRC's financial management systems substantially comply with FFMIA, and (4) complying with applicable laws and regulations. NRC management evaluated the effectiveness of NRC's internal control over financial reporting as of September 30, 2010, based on criteria established under FMFIA. NRC management's assurances are included in the Systems, Controls, and Legal Compliance section of the Management's Discussion and Analysis.

An entity's internal control over financial reporting is a process effected by those charged with governance, management, and other personnel, the objectives of which are to provide reasonable assurance that (1) transactions are properly recorded, processed, and summarized to permit the preparation of financial statements in accordance with U.S. generally accepted accounting principles, and assets are safeguarded against loss from unauthorized acquisition, use, or disposition; and (2) transactions are executed in accordance with the laws governing the use of budget authority and other laws and regulations that could have a direct and material effect on the financial statements.

Objectives, Scope and Methodology

We are responsible for planning and performing our audit to obtain reasonable assurance about whether the financial statements are free of material misstatement. An audit includes examining, on a test basis, evidence supporting the amounts and disclosures in the financial statements. An audit also includes assessing the accounting principles used and significant estimates made by management, as well as evaluating the overall financial statement presentation.

We are responsible for planning and performing our examination to obtain reasonable assurance about whether management maintained, in all material respects, effective internal control over financial reporting as of September 30, 2010. Our examination included obtaining an understanding of NRC and its operations, including internal control over financial reporting; considering NRC's process for evaluating and reporting on internal control over financial reporting which the NRC is required to perform by FMFIA; assessing the risk that a material misstatement exists in the financial statements and the risk that a material weakness exists in internal control over financial reporting; evaluating the design and operating effectiveness of internal control and assessing risk; testing relevant internal controls over financial reporting; and performing such other procedures as we considered necessary in the circumstances. We did not test all internal controls relevant to operating objectives as broadly defined by FMFIA.

Because of inherent limitations in any internal control, misstatements due to error or fraud may occur and not be detected. Also, projections of any evaluation of the internal control to future periods are subject to the risk that the internal control may become inadequate because of changes in conditions, or that the degree of compliance with the policies or procedures may deteriorate.

We are also responsible for testing compliance with selected provisions of laws and regulations that have a direct and material effect on the financial statements. We did not test compliance with all laws and regulations applicable to the NRC. We limited our tests of compliance to those laws and regulations required by OMB audit guidance that we deemed applicable to the financial statements for the fiscal years ended September 30, 2010 and 2009. We caution that noncompliance may occur and not be detected by these tests and that such testing may not be sufficient for other purposes.

We conducted our audit and examinations in accordance with auditing standards generally accepted in the United States of America; Government Auditing Standards, issued by the Comptroller General of the United States; attestation standards established by the American Institute of Certified Public Accountants; and OMB Bulletin No. 07-04, Audit Requirements for Federal Financial Statements, as amended. We believe that our audit and examinations provide a reasonable basis for our opinions.

We noted less significant matters involving the NRC's internal control and its operation, which we have reported to the management of the NRC separately.

Distribution

This report is intended solely for the information and use of the NRC OIG, the management of NRC, OMB, the Government Accountability Office and the Congress of the United States, and is not intended to be and should not be used by anyone other than these specified parties.

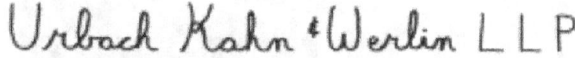

Arlington, Virginia
November 7, 2010

U.S.NRC
UNITED STATES NUCLEAR REGULATORY COMMISSION

Management's Response to the Independent Auditor's Report on the Financial Statements

UNITED STATES
NUCLEAR REGULATORY COMMISSION
WASHINGTON, D.C. 20555-0001

OFFICE OF THE
CHIEF FINANCIAL OFFICER

November 8, 2010

MEMORANDUM TO: Stephen D. Dingbaum
Assistant Inspector General for Audits
Office of the Inspector General

FROM: J. E. Dyer   /RA/
Chief Financial Officer

SUBJECT: AUDIT OF THE FISCAL YEAR 2010 AND 2009 FINANCIAL STATEMENTS

We appreciate the collaborative relationship between the Office of the Inspector General, the auditors, and the Office of the Chief Financial Officer in supporting our continuing effort to improve financial reporting. We have reviewed the Independent Auditor's Report of the Agency's Fiscal Year 2010 and 2009 financial statements and are in agreement with it.

cc: N. Mamish, AO/OEDO
J. Arildsen, OEDO
C. Jaegers, OEDO

Other Accompanying Information

Virgil C. Summer Nuclear Station Site, Jenkinsville, SC.

Photo Courtesy of NRC Photo Library

Desiree Smith and Anne Boland in the Clinton Nuclear Station Incident Response Exercise – October 2009.

Photo Courtesy of NRC Photo Library

Inspector General's Assessment of the Most Serious Management and Performance Challenges Facing the NRC

UNITED STATES
NUCLEAR REGULATORY COMMISSION
WASHINGTON, D.C. 20555-0001

OFFICE OF THE
INSPECTOR GENERAL

October 1, 2010

MEMORANDUM TO: Chairman Jaczko

FROM: Hubert T. Bell /RA/
Inspector General

SUBJECT: INSPECTOR GENERAL'S ASSESSMENT
OF THE MOST SERIOUS MANAGEMENT
AND PERFORMANCE CHALLENGES
FACING NRC (OIG-11-A-01)

The *Reports Consolidation Act of 2000* requires the Inspector General of each Federal agency to annually summarize what he or she considers to be the most serious management and performance challenges facing the agency and to assess the agency's progress in addressing those challenges. In accordance with the act, I identified seven management and performance challenges confronting the Nuclear Regulatory Commission that I consider to be the most serious.

We appreciate the cooperation extended to us during this evaluation. The agency provided comments on this report, which have been incorporated as appropriate. If you have any questions, please contact Stephen D. Dingbaum, Assistant Inspector General for Audits, at 415-5915 or me at 415-5930.

Attachment: As stated

Electronic Distribution

Edwin M. Hackett, Executive Director, Advisory Committee on Reactor Safeguards

E. Roy Hawkens, Chief Administrative Judge, Atomic Safety and Licensing Board Panel

Stephen G. Burns, General Counsel

Brooke D. Poole, Director, Office of Commission Appellate Adjudication

James E. Dyer, Chief Financial Officer

Hubert T. Bell, Inspector General

Margaret M. Doane, Director, Office of International Programs

Rebecca L. Schmidt, Director, Office of Congressional Affairs

Eliot B. Brenner, Director, Office of Public Affairs

Annette Vietti-Cook, Secretary of the Commission

R. William Borchardt, Executive Director for Operations

Michael F. Weber, Deputy Executive Director for Materials, Waste, Research, State, Tribal, and Compliance Programs, OEDO

Darren B. Ash, Deputy Executive Director for Corporate Management, OEDO

Martin J. Virgilio, Deputy Executive Director for Reactor and Preparedness Programs, OEDO

Nader L. Mamish, Assistant for Operations, OEDO

Kathryn O. Greene, Director, Office of Administration

Patrick D. Howard, Director, Computer Security Office

Roy P. Zimmerman, Director, Office of Enforcement

Charles L. Miller, Director, Office of Federal and State Materials and Environmental Management Programs

Cheryl L. McCrary, Director, Office of Investigations

Thomas M. Boyce, Director, Office of Information Services

James F. McDermott, Director, Office of Human Resources

Michael R. Johnson, Director, Office of New Reactors

Catherine Haney, Director, Office of Nuclear Material Safety and Safeguards

Eric J. Leeds, Director, Office of Nuclear Reactor Regulation

Brian W. Sheron, Director, Office of Nuclear Regulatory Research

Corenthis B. Kelley, Director, Office of Small Business and Civil Rights

James T. Wiggins, Director, Office of Nuclear Security and Incident Response

Marc L. Dapas, Acting Regional Administrator, Region I

Luis A. Reyes, Regional Administrator, Region II

Mark A. Satorius, Regional Administrator, Region III

Elmo E. Collins, Jr., Regional Administrator, Region IV

EVALUATION REPORT

Inspector General's Assessment of the
Most Serious Management and Performance
Challenges Facing NRC

OIG-11-A- 01   October 1, 2010

All publicly available OIG reports (including this report) are accessible through
NRC's Web site at:
http:/www.nrc.gov/reading-rm/doc-collections/insp-gen/

EXECUTIVE SUMMARY

BACKGROUND

The *Reports Consolidation Act of 2000* requires the Inspector General (IG) of each Federal agency to annually summarize what he or she considers to be the most serious management and performance challenges facing the agency and to assess the agency's progress in addressing those challenges.

PURPOSE

In accordance with the act, the IG at the U.S. Nuclear Regulatory Commission (NRC) updated what he considers to be the most serious management and performance challenges facing NRC. The IG evaluated the overall work of the Office of the Inspector General (OIG), the OIG staff's general knowledge of agency operations, and other relevant information to develop and update his list of management and performance challenges. As part of the evaluation, OIG staff sought input from NRC's Chairman, Commissioners, and management to obtain their views on what challenges the agency is facing and what efforts the agency has taken to address previously identified management challenges.

RESULTS IN BRIEF

The IG identified seven challenges that he considers the most serious management and performance challenges facing NRC. The challenges identify critical areas or difficult tasks that warrant high-level management attention.

The 2010 list of challenges reflects one change from the 2009 list. Prior Challenge 6, *Administration of all aspects of financial management,* was reworded to include a reference to procurement. The new wording, *Administration of all aspects of financial management and procurement,* is intended to reflect the overarching responsibility that NRC has to manage and exercise stewardship over its resources.

The following chart provides an overview of the seven most serious management and performance challenges as of October 1, 2010.

Most Serious Management and Performance Challenges Facing the Nuclear Regulatory Commission * as of October 1, 2010 (as identified by the Inspector General)	
Challenge 1	*Protection of nuclear material used for civilian purposes.*
Challenge 2	*Managing information to balance security with openness and accountability.*
Challenge 3	*Ability to modify regulatory processes to meet a changing environment, to include the licensing of new nuclear facilities.*
Challenge 4	*Oversight of radiological waste.*
Challenge 5	*Implementation of information technology and information security measures.*
Challenge 6	*Administration of all aspects of financial management and procurement.*
Challenge 7	*Managing human capital.*

*The most serious management and performance challenges are not ranked in any order of importance.

CONCLUSION

The seven challenges contained in this report are distinct, yet interdependent relative to the accomplishment of NRC's mission. For example, the challenge of managing human capital affects all other management and performance challenges.

The agency's continued progress in taking actions to address the challenges presented should facilitate achieving the agency's mission and goals.

ABBREVIATIONS AND ACRONYMS

CFR	Code of Federal Regulations
CUI	Controlled Unclassified Information
FY	fiscal year
IG	Inspector General
HSPD-12	Homeland Security Presidential Directive-12
IMPEP	Integrated Materials Performance Evaluation Program
NMMSS	Nuclear Materials Management and Safeguards System
NRC	U.S. Nuclear Regulatory Commission
NSTS	National Source Tracking System
OIG	Office of the Inspector General
3WFN	Three White Flint North

TABLE OF CONTENTS

I. BACKGROUND

On January 24, 2000, Congress enacted the *Reports Consolidation Act of 2000* (Reports Act), requiring Federal agencies to provide financial and performance management information in a more meaningful and useful format for Congress, the President, and the public. The Reports Act requires the Inspector General (IG) of each Federal agency to annually summarize what he or she considers to be the most serious management and performance challenges facing the agency and to assess the agency's progress in addressing those challenges.

II. PURPOSE

In accordance with the Reports Act's provisions, the U.S. Nuclear Regulatory Commission (NRC) IG updated what he considers to be the most serious management and performance challenges facing the agency. The IG evaluated the overall work of the Office of the Inspector General (OIG), the OIG staff's general knowledge of agency operations, and other relevant information to develop and update his list of management and performance challenges.

In addition, OIG sought input from NRC's Chairman, Commissioners, and management to obtain their views on what challenges the agency is facing and what efforts the agency has taken or planned to address previously identified management and performance challenges.

III. EVALUATION RESULTS

The NRC's mission is to license and regulate the Nation's civilian use of byproduct, source, and special nuclear materials to ensure adequate protection of public health and safety, promote the common defense and security, and protect the environment. Like other Federal agencies, NRC faces management and performance challenges in carrying out its mission.

Determination of Management and Performance Challenges

Congress left the determination and threshold of what constitutes a most serious management and performance challenge to the discretion of the Inspectors General. As a result, the IG applied the following definition in identifying challenges:

> Serious management and performance challenges are mission critical areas or programs that have the potential for a perennial weakness or vulnerability that, without substantial management attention, would seriously impact agency operations or strategic goals.

Based on this definition, in 2010, the IG assessed the most serious management and performance challenges facing NRC and identified seven challenges that he considered most serious. The challenges identify critical areas or difficult tasks that warrant high-level management attention. The 2010 list of challenges reflects one change from the 2009 list:

- Prior Challenge 6, *Administration of all aspects of financial management,* was reworded to include a reference to procurement. The new wording, *Administration of all aspects of financial management and procurement,* is intended to reflect the overarching responsibility that NRC has to manage and exercise stewardship over its resources.

The following chart provides an overview of the seven challenges identified as most serious. The sections that follow the chart provide more detailed descriptions of the challenges, descriptive examples related to the challenges, and examples of efforts that the agency has taken or are underway or planned to address the challenges.

Most Serious Management and Performance Challenges Facing the Nuclear Regulatory Commission * as of October 1, 2010 (as identified by the Inspector General)	
Challenge 1	Protection of nuclear material used for civilian purposes.
Challenge 2	Managing information to balance security with openness and accountability.
Challenge 3	Ability to modify regulatory processes to meet a changing environment, to include the licensing of new nuclear facilities.
Challenge 4	Oversight of radiological waste.
Challenge 5	Implementation of information technology and information security measures.
Challenge 6	Administration of all aspects of financial management and procurement.
Challenge 7	Managing human capital.

The most serious management and performance challenges are not ranked in any order of importance.

CHALLENGE 1
Protection of nuclear material used for civilian purposes.

NRC is authorized to grant licenses for the possession and use of radioactive materials and establish regulations to govern the possession and use of those materials.

NRC's regulations require that certain material licensees have extensive material control and accounting programs as a condition of their licenses. All other license applications (including those requesting authorization to possess small quantities of special nuclear materials) must develop and implement plans that demonstrate a commitment to accurately control and account for radioactive materials.

NRC may relinquish to States, upon their request, its authority to regulate certain radioactive materials and limited quantities of special nuclear material. After these States demonstrate that their regulatory programs are adequate to protect public health and safety and compatible with NRC's program, the States enter into an agreement assuming this regulatory authority from NRC and are called Agreement States.

The issues related to this challenge and the agency's actions to address each issue include the following:

Issue: Implement the National Source Tracking System (NSTS), Web Based Licensing, and the Licensing Verification System to ensure the accurate tracking and control of byproduct material, especially those materials with the greatest potential to impact public health and safety.

> **Action:** NSTS became operational in December 2008 and was available to licensees in January 2009 for tracking Code of Conduct[1] materials in categories 1 and 2. Although there are some issues regarding NSTS' credentialing process, the staff has

---

[1] In January 2004, the International Atomic Energy Agency published the Code of Conduct on the Safety and Security of Radioactive Sources as the standard the international community uses to govern the safety and security of radioactive materials based on the categorization system. While the International Atomic Energy Agency classifies sources into five categories, it notes that sources in categories one through three are designated as varying degrees of dangerous.

numerous actions underway to address these difficulties and ultimately increase the online usage of NSTS. The agency is still working to make operational Web Based Licensing and the Licensing Verification System. The agency recently awarded the *Integrated Source Management Portfolio* contract, which will integrate NSTS, Web Based Licensing, and the Licensing Verification System to license and track source materials under one mechanism.

Issue: Ensure that radioactive material is adequately protected to preclude its use for malicious purposes.

> **Action:** Although NRC initiated a rulemaking to expand the materials tracked in NSTS, the decision and potential implementation of that rulemaking was not approved by the Commission. A Commissioner highlighted the following improvements in NRC's overall licensing process as a reason for not approving the rulemaking: background investigations; increased inspections; additional license review; pre-licensing verification and site visits; transfer of sources under existing security orders to verify new users; flagging of significant changes in ordering patterns; licensing of end users; requirements in Title 10, Code of Federal Regulations (CFR), Part 30.41, requiring licensees to verify that a recipient is authorized to receive material; and the presence of existing increased control orders for licensees possessing quantities of material that in the aggregate exceed Category 2 levels.
>
> One radioactive isotope that is of particular concern for malicious use is cesium-chloride. The U.S. National Academy of Sciences issued a report that emphasized replacement technologies be considered for cesium-chloride, a highly dispersible chemical form of the radioactive isotope of Cesium, Cs-137. Cesium-chloride is very soluble in water and easily dispersed in the air and is highly toxic if ingested. Cesium-chloride, used in nuclear medicine, research, and industry, is typically double sealed and contained in a stainless steel capsule for safety reasons. In light of the views on alternative technologies as a replacement, NRC convened public workshops to seek input from various stakeholders. NRC also commissioned a study by its *Advisory Committee on the Medical*

Uses of Isotopes. After carefully considering all these inputs, as well as the NRC's own internal analysis, the agency concluded that near-term replacement of cesium-chloride devices was not practicable, and would be detrimental to the delivery of medical care and research. The current policy allows the continued use of cesium chloride while actively pursuing a better alternative. Additionally, NRC issued a draft policy statement for public comment that emphasizes that developing alternatives to cesium-chloride sources would be prudent.

Issue: Ensure the appropriate oversight of uranium recovery facilities.

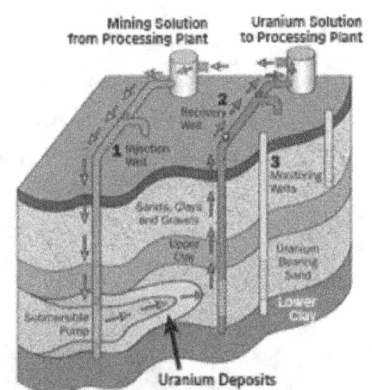

Action: NRC maintains a regulatory oversight program with respect to licensing and inspection of uranium recovery facilities to ensure that licensees conduct activities safely and in an environmentally protective manner. NRC regulates four in situ recovery facilities[2] in the Western States, and, for those operating facilities, conducts routine annual inspections to ensure that they are safely operated.

Issue: Ensure adequate inspections to verify licensees' commitments to their material control and accounting programs.

Action: NRC is enhancing its inspection program. Currently, fuel cycle material control and accounting inspections are a shared responsibility between the Office of Nuclear Material Safety and Safeguards and NRC's Region II. The agency continues to ensure that there are two material control and accounting inspectors in each location.

Additionally, NRC is working to document the basis for risk-informing its material control and accounting program with respect to conducting periodic inspections.

[2] In situ recovery is one of the two primary extraction methods that are currently used to obtain uranium from underground. These facilities recover uranium from low-grade ores where other mining and milling methods may be too expensive or environmentally disruptive.

U.S.NRC

UNITED STATES NUCLEAR REGULATORY COMMISSION

Issue: Ensure reliable accounting of special nuclear materials in the NRC and Department of Energy's jointly managed Nuclear Materials Management and Safeguards System (NMMSS).

Action: NRC has been working since 2003 to resolve issues of material control and accounting in response to OIG-03-A-15, *Audit of NRC's Regulatory Oversight of Special Nuclear Materials*. On February 7, 2008, NRC approved a final rule that amended its regulations to improve the accuracy of material inventory information maintained in the NMMSS. The amendments, effective January 1, 2009, lower the threshold of reportable quantities of special nuclear materials and certain source materials to the NMMSS, modify the types and timing of submittals to the NMMSS, and require licensees to reconcile any material inventory discrepancies that NRC identifies in the NMMSS database. NRC reports that it has started implementing the rule change requiring improved reporting and reconciliation for licensees reporting to NMMSS, and has verified the adequacy of material control and accounting of special nuclear material at NRC licensed facilities. Additionally, the Commission has directed the NRC staff to revise and consolidate current material control and accounting regulations into 10 CFR Part 74. This final rule and associated guidance is scheduled to be completed by April 30, 2012.

Issue: Ensure that Agreement State programs are adequate to protect public health and safety and the environment, and are compatible with NRC's program.

Action: NRC conducts 8 to 10 reviews per year of Agreement State radioactive materials programs and NRC's regional programs under the agency's *Integrated Materials Performance Evaluation Program* (IMPEP). Furthermore, NRC completed a self-assessment of IMPEP in July 2010. To date, the agency has evaluated the IMPEP Team Member Training and audited the preparations and onsite portion of an IMPEP review. NRC also plans to review IMPEP policies and procedures, interview agency and Agreement State managers and staff, and develop a procedure for future self-assessments.

CHALLENGE 2
Managing information to balance security with openness and accountability.

NRC employees create and work with a significant amount of sensitive information that needs to be protected. Such information includes sensitive unclassified information and classified national security information contained in written documents and various electronic databases.

In addressing continuing terrorist activity worldwide, NRC continually reexamines its information management policies and procedures. NRC faces the challenge of attempting to balance the need to protect sensitive information from inappropriate disclosure with the agency's goal of openness in its regulatory processes. Over the past year, NRC has made various efforts to improve public access to information while protecting sensitive information, including security-related information, from inappropriate disclosure.

The issues related to this challenge and the agency's actions to address each issue include the following:

<u>Issue</u>: Be responsive to requests for information and provide external stakeholders with clear and accurate information about regulatory programs and facilitate public participation in the regulatory process to ensure openness and accountability.

> <u>**Action:**</u> NRC instituted a contract to review documents that were removed from the public domain after September 11, 2001, and restore them to the public domain, in their entirety or redacted, whichever is appropriate. This contract will remain in effect through fiscal year (FY) 2011.

> <u>**Action:**</u> NRC continues to evaluate security related information to determine what can be made publicly available. Information that would not be beneficial to an adversary has been made available to the public through various means including the NRC Web site, a public version of the annual security report to Congress, and public meetings. A recent example is a series of public meetings addressing proposed improvements to the significance determination process.

Action: NRC staff have conducted a number of stakeholder outreach efforts to include public meetings on specific regulatory issues and with elected officials regarding issues at facilities within their jurisdiction.

Action: In response to recommendations in OIG's *Audit of NRC's Process for Closed Meetings* (OIG-A-14), the agency is planning to implement several measures to better notify the public about when NRC holds non-public meetings with external stakeholders and to what topics these meetings pertain.

Issue: Manage information in accordance with new Federal Government policies for designating, marking, safeguarding, and disseminating controlled unclassified information (CUI).

Action: In May 2009, the President issued a memorandum on CUI which established an interagency Task Force to review the CUI framework. Following further presidential direction, the National Archives and Records Administration will issue implementation guidance and NRC will develop its plan to implement the new CUI program. NRC's Safeguards Information program will be incorporated into the CUI program.

Issue: Ensure that sensitive information is handled in accordance with agency policies and procedures for public disclosure.

Action: NRC announced the release of the "NRC's Personally Identifiable Information Responsibilities Awareness and Acknowledgement of Understanding" training presentation through Yellow Announcement No. 116 dated November 16, 2009. NRC developed this presentation to ensure that all personnel are aware of their responsibilities for protecting Personally Identifiable Information, understand the consequences of violating of these responsibilities, and acknowledge these responsibilities on an annual basis.

Action: In addition, NRC is in the process of reviewing the shared network drives to ensure that Personally Identifiable Information is adequately protected or removed if unnecessary.

Issue: Review and strengthen programs to protect licensee, vendor, and Government-owned assets (e.g., facility designs, technology descriptions, dual use material and components, classified information) from compromise by foreign sources and industrial espionage and increase awareness of the relationship of these assets to the Nation's economic and industrial base and energy infrastructure.

> **Action:** NRC has recognized the need to ensure technological data involving licensee, vendor, and Government-owned assets is fully protected against potential loss to adversaries. NRC has promulgated orders that provide additional security measures for the protection of these assets.
>
> NRC employees and contractors are required to have a baseline level of security awareness upon entry on duty and the receipt of a security clearance. Others, depending on their job and involvement in the creation and use of protected information, are provided various "role based" training programs, such as classifier's training, training for administrative personnel, declassification training, Secret Internet Protocol Router Network users training, and Sensitive Compartmented Information Access training. The training is layered, targeted, and recurring for those who have specific responsibilities for various types of protected information.
>
> In addition, NRC has increased its information security awareness through the issuance of a variety of agencywide announcements informing staff of the methods employed by those targeting NRC information systems and the corresponding need for employees to heighten their computer security information protection posture.

Issue: Technologies or materials, which the NRC regulates, have potential intelligence value to foreign states and non-state actors from either an intelligence or a counterproliferation, counterterrorism, or economic espionage perspective and should be protected from potential compromise. Further, there is the potential that NRC employees have knowledge and access to information that may be of interest to foreign powers and non-state actors.

> **Action:** The NRC has begun the process of developing programmatic efforts aimed at identifying potential threats and vulnerabilities that exist in its programs and operations. Such efforts should continue and receive senior leadership support.

> ### CHALLENGE 3
> *Ability to modify regulatory processes to meet a changing environment, to include the licensing of new nuclear facilities.*

NRC faces the challenge of maintaining its core regulatory programs while adapting to changes in its regulatory environment. NRC must address a growing interest in licensing and constructing new nuclear power plants to meet the Nation's increasing demands for energy production. As of June 2010, NRC had received 18 Combined Operating License applications, and the agency expects to receive 2 new applications through FY 2012.

While responding to the emerging demands associated with licensing and regulating new reactors, NRC must maintain focus and effectively carry out its current regulatory responsibilities, such as inspections of the current fleet of operating nuclear reactors and fuel cycle facilities. NRC intends to increase its safety focus on licensing and oversight activities through risk-informed and performance-based regulation.

The issues related to this challenge and the agency's actions to address each issue include the following:

New Facilities

Issue: Implement the new Construction Inspection Program.

• Risk-inform Construction Inspection Program activities to ensure the safe operation of newly constructed nuclear facilities.

• Ensure that the NRC staff has the necessary knowledge and skill to successfully implement the program.

> **Action:** The Office of New Reactors has developed the new Construction Inspection Program in accordance with 10 CFR Part 52. New inspections, tests, analyses, and acceptance criteria have been integrated into the Part 52 licensing process "to create a design-specific, pre-approved set of performance standards that the licensee must meet and that the Commission must find have been met, before the licensee can load fuel and operate the plant." Additionally, the agency has issued and revised a number of Inspection Manual Chapters and procedures to

implement the new inspections, tests, analyses, and acceptance criteria process.

The Office of New Reactors continues to make improvements to its construction inspection and quality assurance practices per OIG recommendations. For example, NRC has revised Inspection Manual Chapter 1252, *Construction Inspector Training and Qualification Program,* to ensure that the agency is effectively preparing inspectors to implement the new Construction Inspection Program. The agency will monitor the effectiveness of the training program as inspections of the new construction projects begin.

Issue: As the public's demand for new energy sources continues, NRC must ensure that the process for reviewing applications for new facilities focuses on safety and effectiveness.

Action: NRC's preparations have been focused on issuing reactor design certifications, revising the regulation that governs early site permits, and engaging in ongoing interactions with nuclear plant designers and utilities regarding prospective new reactor applications and licensing activities. In April 2009, the Office of New Reactors developed a set of goals with the purpose of enhancing the agency's ability to plan and implement its reviews more effectively in a dynamic environment resulting from changes in the applicants' business strategies.

NRC is taking a "design-centered review approach" to optimize the Combined Operating License application review process. Part of the license review process includes conducting risk-informed performance-based vendor inspections and quality assurance/quality control audits.

Issue: As the sources of manufactured reactor components become more globalized, NRC must ensure its regulations and oversight activities appropriately address the challenges associated with licensees procuring components from suppliers located outside of the United States.

Action: The Office of New Reactors has taken steps to allocate resources for the use of translators and/or interpreters to support the office's foreign vendor inspections. NRC also participates in the Multinational Design Evaluation Program, which is a multinational

initiative taken by national safety authorities to develop approaches to leverage the resources and knowledge of the national regulatory authorities who will be tasked with the review of new reactor power plant designs.

Existing Fleet

Issue: Ensure NRC maintains the ability to effectively review licensee applications for license renewals and power uprates submitted by industry in response to the Nation's increasing demands for energy production.

> **Action:** For planning purposes, NRC continues to work with plant licensees to develop a schedule of anticipated license amendment requests for license renewals and power uprates. The agency has also implemented a number of recommendations to improve the license renewal review and power uprate processes to include closer management oversight. For license renewal reviews, the agency has updated report-writing guidance to include management expectations and report-writing standards. For power uprate reviews, the agency has developed a training module for technical reviewers and project managers that is specifically focused on writing or contributing to a safety evaluation.

Issue: Respond to a heightened public focus on license renewals resulting in contested hearings.

> **Action:** NRC has open dialogs with the industry, licensees, and stakeholders, and appropriate comments have been incorporated into new inspection procedures. Additionally, the license renewal process allows stakeholders to request a hearing in order to present their concerns.

Issue: Ensure the ability to identify emerging operating and safety issues at all plants, including issues associated with license renewal and power uprate; consistently apply regulatory and review changes in response to these emerging issues across the existing fleet of reactors.

> **Action:** NRC continues to make changes to its regulatory programs based on emerging operational and safety issues related to license renewal and power uprate. For example, as a result of identified

weaknesses in the power uprate program, Inspection Procedure 71004 was revised to provide additional guidance on inspection planning, implementation, and documentation. Annually, agency staff communicate the status of the license renewal and power uprate programs to the Commission.

In March 2010, NRC formed a Groundwater Contamination Task Force to review the actions taken in response to recent releases of tritium into groundwater by nuclear facilities. In June 2010, the Task Force issued a report with 16 conclusions and 4 specific recommendations for the agency to strengthen NRC's response to groundwater incidents.

Issue: Establish and maintain effective, stable, and predictable regulatory programs or policies for all programs.

Action: NRC continues to interface with stakeholders, develop regulatory policy, update rules and technical guidance, provide technical lead and management for the Reactor Oversight Process, and support the development of programmatic changes when needed. Additionally, the Reactor Oversight Process features an annual assessment process which is used to revise the program as necessary.

![U.S.NRC logo] **U.S.NRC**
UNITED STATES NUCLEAR REGULATORY COMMISSION

CHALLENGE 4
Oversight of radiological waste.

NRC regulates spent nuclear fuel generated from commercial nuclear power reactors, which is referred to as high-level radioactive waste. NRC faces significant issues involving the uncertainty of a potential withdrawal of a Department of Energy license application for the Yucca Mountain repository for storing high-level radioactive waste. Additional challenges in the high-level waste area include the interim storage of spent nuclear fuel, certification of storage and transportation casks, and the oversight of decommissioned reactors and other nuclear sites.

Additionally, the amount of low-level waste continues to grow; however, no new disposal facilities have been built since the 1980s, and unresolved issues will increase as access to disposal facilities becomes more limited given facility closures and restricted accessibility.

The issues related to this challenge and the agency's actions to address each issue include the following:

Issue: Address increasing quantities of radiological waste requiring interim storage or permanent disposal.

Action: NRC developed and implemented a risk-informed decisionmaking framework in connection with a wide range of nuclear waste storage issues. NRC has conducted reviews using the framework for dry cask waste storage systems and concluded that such systems provide a safe means to store spent nuclear fuel with exceedingly low risk. NRC has met with Agreement States and issued guidance on interim storage of low-level nuclear waste. Stakeholder outreach is an integral part of the implementation of NRC's low level waste strategic assessment.

Issue: Address issues regarding the uncertainty of NRC's continued review or the potential withdrawal of the Department of Energy's license application to construct a high-level radioactive waste repository at Yucca Mountain.

Action: NRC is continuing to review the Yucca Mountain license application submitted by the Department of Energy in June 2008. In

2010, the agency held hearings and evaluated a wide range of technical and scientific issues. On August 2, 2010, the agency issued Volume 1 of a safety evaluation report on the U.S. Department of Energy license application to construct a geologic repository at Yucca Mountain, Nevada. Volume 1 contains the NRC staff's conclusion that the "General Information" section of the Department of Energy license application adequately describes the proposed repository. A final decision on the application will be made after completion of NRC's independent technical review of the application, an adjudicatory hearing, and subsequent Commission review. If the Department of Energy successfully withdraws its license application for a high-level waste repository, NRC staff plan to conduct an orderly shutdown of the technical review program, including knowledge management and responding to continued intervener appeals.

Issue: Oversight of low-level waste storage and disposal, including low-level radioactive waste disposal sites. All current low-level waste disposal sites are regulated by Agreement States.

> **Action:** NRC has focused on stakeholder outreach as an integral part of the Low-Level Waste Strategic Assessment. This outreach communicates to licensees that the NRC's staff position continues to be that low-level waste storage must meet NRC requirements to ensure safe operation, and that when constructing new low-level waste storage facilities, the regulations for evaluating proposed changes to facilities must be met. In March 2010, NRC posted to its Web site guidance on long-term storage of low-level waste.

Issue: Oversight of nuclear waste issues associated with the decommissioning and cleanup of nuclear reactor sites and other facilities.

> **Action:** NRC continues to hold public meetings with stakeholders and licensees to explore safe and secure storage options associated with decommissioning of plants, such as transitioning from spent pool storage to dry cask storage. NRC continues to oversee the 13 power reactors currently undergoing decommissioning. NRC staff published NUREG-1307, "Report on Waste Burial Charges," which provides updated disposal costs for pressurized water reactors and boiling water reactors based on estimated disposal volumes.

CHALLENGE 5
Implementation of information technology and information security measures.

NRC needs to continue upgrading and modernizing its information technology and security capabilities both for employees and for public access to the regulatory process. Recognizing the need to modernize, the Office of Information Services established goals to improve the productivity, efficiency, and effectiveness of agency programs and operations, and enhance the use of information for all users inside and outside the agency. NRC also needs to ensure that system security controls are in place to protect the agency's information systems against misuse.

The issues related to this challenge and the agency's actions to address each issue include the following:

Issue: Upgrade and manage information technology activities to improve the productivity, efficiency, and effectiveness of agency programs and operations.

> **Action**: An aggressive implementation schedule was developed to upgrade the existing information technology environment and to bring new technologies to NRC. Projects under development include a virtual private network and standard laptop and dockable workstation configuration. The Computer Security Office has also established mandatory laptop security standards, including requirements for full-disk encryption and security wireless capabilities for users outside of the NRC network. These efforts, which were underway in FY 2010 and will continue during FY 2011, are intended to enable NRC staff to securely access and use the systems and information needed to perform job junctions, regardless of where they are located.

> **Action**: To further agency plans for technology modernization, the Office of Information Services began analysis of information technology/information management legacy applications with business owners to identify opportunities for transforming legacy applications starting in FY 2013. The office continues to work with offices to develop a funding strategy for application modernization. In addition, the office now offers business analysis services with the goal of improving requirements definition.

Action: The Office of Information Services has developed an enterprise contracting strategy for commonly used information technology services to improve productivity and efficiency in information technology contracting.

Action: NRC Implemented Microsoft Office 2007, thereby upgrading the office suite of applications to a current platform. The agency also implemented Internet Explorer 8 to upgrade the current Web browser to a current and more secure application.

Issue: Provide laptop computers with enhanced functionality, security, and support.

Action: The agency has set goals concerning laptops for the Office of Information Services to implement over the next several years. The agency has identified and is addressing its needs to (1) develop policies and standards for the use of laptop computers, (2) implement enterprise encryption and updating of operating systems to support the laptop program, and (3) provide secure wireless capability access.

Issue: Ensure that information systems and assets are protected.

Action: The Computer Security Office has taken action on identified vulnerabilities. Such actions include (1) certifying and accrediting all general support systems and major applications that are reported to the Office of Management and Budget in accordance with the Federal Information Security Management Act; (2) initiating a continuous monitoring process to annually evaluate information technology security controls of agency information technology systems to provide assurance that systems remain secure after having been authorized to operate; (3) publishing information technology security policy and standards to address current agency needs; and (4) implementing a security impact assessment process for evaluating the nature and extent of changes to information technology systems that have been authorized to operate.

Action: The NRC is deploying a variety of capabilities that strengthen its ability to identify, mitigate and ameliorate threats against its information systems infrastructure. These means, coupled with cyber tabletop exercises designed to examine the agency's response to

potential network intrusion attacks, provide the NRC with enhanced capabilities to respond to such threats.

Action: The agency has established a secure network that enables authorized users to access safeguards information documents electronically. This system will reduce the need to print documents and will enable the management of safeguards documents in a centralized electronic document management system.

Action: The agency has issued Homeland Security Presidential Directive-12 (HSPD-12) identification cards to NRC staff and contractors and is working to install HSPD-12 card readers at headquarters and regional facilities. Use of this technology is expected to reduce the risk of unauthorized personnel gaining access to NRC facilities, thereby improving security of sensitive information and information technology assets.

Issue: Ensure that plans for a cyber security inspection program are developed and implemented.

Action: The staff plans to develop an inspection procedure for conducting cyber security inspections at nuclear power plants and hold training for NRC cyber security inspectors. The inspections are planned to be conducted between calendar years 2012 and 2016.

CHALLENGE 6

Administration of all aspects of financial management and procurement.

NRC management is also responsible to meet the objectives of several statutes, including the Federal Managers' Financial Integrity Act. This act mandates that NRC establish controls that reasonably ensure that (1) obligations and costs comply with applicable law; (2) assets are safeguarded against waste, loss, unauthorized use, or misappropriation; and (3) revenues and expenditures are properly recorded and accounted for. This act encompasses programmatic and administrative areas, as well as accounting and financial management.

NRC's procurement of goods and services must be made with an aim to achieve the best value for the agency's dollars in a timely manner. Further, agency policy provides that NRC's procurement of goods and services supports the agency's mission; is planned, awarded, and administered efficiently and effectively; and is consistent with sound business practices and contracting principles.

The issues related to this challenge and the agency's actions to address each issue include the following:

Financial Management

Issue: Replace the agency's current financial systems, which are obsolete, overly complex, and inefficient.

> **Action**: The agency is scheduled to deploy the new Financial Accounting and Integrated Management Information System (FAIMIS) on October 1, 2010. The FAIMIS core implementation replaces the functionality of five core financial systems with a single Web-based system based on a commercial-off-the-shelf software system. The agency plans to deploy an acquisitions module for FAIMIS in October 2012.

> **Action**: NRC's plans to upgrade the Time and Labor System in July 2010 were delayed because of performance issues identified during production testing. The agency is currently analyzing the performance

issues to determine the root causes. Once the root causes have been identified, the agency plans to develop a path forward, including a project plan to address implementation strategy, resource requirements, and milestones. NRC will continue to use the legacy system until the new system is deployed.

Action: In July 2009, NRC implemented e-Travel, a Government-wide initiative to improve travel operations and management. The agency expanded implementation of the system in FY 2010 to include specialized travel, such as foreign and premium class travel and split pay, an option that allows employees to apply a portion of their travel reimbursement to pay their Government credit card bill and a portion to their bank account.

Issue: Respond to Commission direction and implement recommendations of the Advisory Group on Budget Formulation and Financial Plan Reporting. This issue encompasses both budget formulation and budget execution.

Action: NRC made improvements in the budget formulation and execution processes consistent with Commission direction and the recommendations of the Advisory Group on Budget Formulation and Financial Plan Reporting.

Budget Formulation: For the FY 2011 budget, the budget formulation process was streamlined and took advantage of the upgraded Budget Formulation System accessibility and functionality enhancements. The formulation of the FY 2011 budget included a new budget structure that incorporates products and product lines.

Budget Execution: During FY 2010, NRC implemented improvements to the budget execution process.

- The midyear review and request process was eliminated and replaced with a reprogramming strategy early in the fiscal year. This resulted in funds being made available earlier in the year. Plans are to continue to streamline and accelerate this process for the next execution year.

- Additionally, advance procurement planning was coordinated with the planning for funds utilization, which brought the agency a step closer to ultimately integrating the advanced procurement plan and budget execution.

- Finally, the new budget structure positions the agency to integrate budget formulation, execution, and performance information using the new financial management systems. Budget execution and management information reporting will be improved through enhanced capabilities to compare budgeted amounts with actual funds used.

Procurement

Issue: Implement improvements in the agency's procedures for awarding, negotiating, and managing agreements with Department of Energy laboratories.

> **Action**: In response to an OIG audit,[3] NRC has agreed to revise Management Directive 11.7, *NRC Procedures for Placement and Monitoring of Work with the U.S. Department of Energy*. The revisions include the following: requiring NRC offices to consider the use of commercial sources through market research, clarifying Management Directive 11.7 to emphasize the requirement to document the rationale and basis for using a Department of Energy lab, and requiring independent review of justifications by NRC Division of Contracts personnel to ensure that commercial sources are fully considered. In addition, the NRC plans to initiate efforts with the Department of Energy to update the memorandum of understanding between the two agencies to require that the Department of Energy provide NRC with timely audit reports.

Issue: Manage the agency's expanded grant program to ensure that grants are awarded in a timely manner and NRC personnel who award and administer grants are provided appropriate training.

---

[3] OIG-10-A-12, *Audit of NRC's Management of Agreements with Department of Energy Laboratories* (April 23, 2010).

Action: NRC established and documented a process for announcing grants, reviewing applications, and administering grants. NRC also implemented the Department of the Treasury's Automated Standard Application for Payments System to ensure accessible and timely distribution of funds to the grantees. This allows funds to be available as early as the beginning of the grant's period of performance for immediate drawdown based on incurred costs.

Action: NRC conducted a Lean Six Sigma[4] review of the agency's process for awarding grants to reduce the overall time for processing grants. Recently, the Division of Contracts issued Interim guidance to offices responsible for implementing the Grants Program to include the Office of Human Resources, Office of Small Business and Civil Rights, and Office of Nuclear Regulatory Research, as recommended by the Lean Six Sigma and by OIG's *Audit of NRC's Grant Management Program.*[5]

Action: On June 24, 2010, the Executive Director for Operations issued a memorandum to the Office of Human Resources, Office of Small Business and Civil Rights, and Office of Nuclear Regulatory Research, establishing a Grants Management Certification and Training Program, effective immediately. The training program will ensure that grants specialists or grants project officers are appropriately trained and certified to carry out their fiduciary responsibilities. The program mandates specific training for staff involved in awarding, administering, and monitoring grants and cooperative agreements. Implementation of this program also responds to recommendations in OIG's *Audit of NRC's Grants Management Program.*

---

[4] Lean Six Sigma is a structured methodology that NRC uses to accomplish sustained improvements to the types of process, transactions, and services that are performed routinely at the agency.
[5] OIG-09-A-16, *Audit of NRC's Grant Management Program* (September 29, 2009).

CHALLENGE 7

Managing human capital.

Over the last 6 years, NRC's workforce has grown from 3,059 staff to approximately 4,000 staff currently. This represents an increase of approximately 33 percent. Some offices still have a need for additional staff to deal with the increased workload in the Low-Level Waste and Uranium Recovery Programs while other offices may face a decreased need resulting from various states becoming Agreement States. To effectively manage human capital, while continuing to accomplish the agency's mission, NRC must continue to implement initiatives in the following areas:

- Recruitment and training.

- Space planning.

The issues related to this challenge and the agency's actions to address each issue include the following:

<u>Issue</u>: NRC must address recruitment, training, and knowledge management in light of anticipated fluctuations in workload demands and retirements.

> <u>Action</u>: NRC is refining the agency's human capital program through the following initiatives: (1) reviewing existing recruitment strategies to determine how the agency can maximize and leverage limited resources to position the agency to be successful with both current and long-term human capital needs; (2) developing a talent acquisition and recruitment plan that will focus on strengthening its academic linkages, diversity, and other areas; and (3) acquiring wellness services to provide for functions involving health, physical fitness, ergonomics, automated external defibrillation, and occupational safety and health services.

> <u>Action</u>: For FY 2011, NRC will strategically focus on fine tuning available skill sets to meet future mission needs. The agency anticipates various critical skill needs for the next several years and

U.S.NRC
UNITED STATES NUCLEAR REGULATORY COMMISSION

will continue to recruit, hire, and develop staff to meet these skill needs. Hiring strategies will also include emphasis on governmentwide programs, specifically hiring of the disabled and employment of veterans.

<u>Issue</u>: NRC needs to facilitate continuation of its space planning efforts. Last October, the General Services Administration signed a lease for the construction and occupation of a building that the developer will construct across Marinelli Road from One White Flint North. NRC will occupy the building, referred to as Three White Flint North (3WFN), under the terms of a 15-year lease between the General Services Administration and the building owners. Ground-breaking ceremonies for the building were held during May with excavation beginning in early July. When completed, 3WFN will provide office space for approximately 1,300 NRC staff members and allow the agency to reconsolidate headquarters staff who are now dispersed among four offsite locations. The space in 3WFN will allow the agency to decompress work areas and restore conference rooms that had been converted to workstations in both One White Flint North and Two White Flint North. The building will also house the Headquarters Operations Center and the Data Center. At the present time, there is no funding in the budget for either above ground or underground pedestrian access between One White Flint North and 3WFN. To access 3WFN, agency employees will have to cross Marinelli Road, which is a multi-lane road. NRC faces two challenges related to 3WFN. The agency must ensure that:

- Building requirements are met and within budget.

- Provisions are put in place to ensure safe pedestrian movement between the buildings.

    <u>Action</u>: Although the construction is in the preliminary phase and the target date to begin moving staff into 3WFN is September 2012, NRC is currently working on the design requirements for the interior of the building. The goal is to provide a good working environment for NRC employees within the budget.

    <u>Action</u>: In late August, the agency signed a Memorandum of Understanding with Montgomery County to ensure cooperation to maximize pedestrian safety around the White Flint complex. In the near

term, the county Department of Transportation staff are meeting with
NRC staff to discuss current and future pedestrian and vehicle flows on
both sides of Marinelli Road and how best to manage them between
Rockville Pike and the White Flint complex vehicle entrance/exit.

IV.   CONCLUSION

The seven challenges contained in this report are distinct, yet are
interdependent to accomplishing NRC's mission. For example, the
challenge of managing human capital affects all other management and
performance challenges.

The agency's continued progress in taking actions to address the
challenges presented should facilitate achieving the agency's mission and
goals.

SCOPE AND METHODOLOGY

This evaluation focused on the IG's annual assessment of the most serious management and performance challenges facing the NRC. The challenges represent critical areas or difficult tasks that warrant high level management attention. To accomplish this work, the OIG focused on determining (1) current challenges, (2) the agency's efforts to address the challenges during FY 2009, and (3) future agency efforts to address the challenges.

OIG reviewed and analyzed pertinent laws and authoritative guidance, agency documents, and OIG reports, and sought input from NRC officials concerning agency accomplishments relative to the challenge areas and suggestions they had for updating the challenges. Specifically, because challenges affect mission critical areas or programs that have the potential to impact agency operations or strategic goals, NRC Commission members, offices that report to the Commission, the Executive Director for Operations, and the Chief Financial Officer were afforded the opportunity to share any information and insights on this subject.

OIG conducted this evaluation from May through August 2010 at NRC Headquarters. The major contributors to this report were Steven Zane, Deputy Assistant Inspector General for Audits; Sherri Miotla, Team Leader; Beth Serepca, Team Leader; Kathleen Stetson, Team Leader; RK Wild, Team Leader; and Judy Gordon, Quality Assurance Manager.

Management Decisions
and Final Actions
on OIG Audit
Recommendations

Photo Courtesy of NRC Photo Library

Monticello Nuclear Power Plant, Monticello, MN.

Photo Courtesy of NRC Photo Library

Resident Inspector's Erin Bonney (left) and Tracey Zeiv (right) assessing the overall condition of the Beaver Valley Power Station, Unit 1 facility, following an 18-month operating cycle – October 2009.

The agency has established and continues to maintain an excellent record in resolving and implementing audit recommendations presented in OIG reports. Section 5(b) of the Inspector General Act of 1978, as amended, requires agencies to report on final actions taken on OIG audit recommendations. The following table gives the dollar value of disallowed costs determined through contract audits conducted by the Defense Contract Audit Agency and NRC's Office of the Inspector General. Because of the sensitivity of contractual negotiations, details of these contract audits are not furnished as part of this report. As of September 30, 2010, there were no outstanding audits recommending that funds be put to better use.

Management Report On Office Of The Inspector General Audits With Disallowed Costs

For the period October 1, 2009 – September 30, 2010

Category	Number of Audit Reports	Questioned Costs	Unsupported Costs
1. Audit reports with management decisions on which final action had not been taken at the beginning of this reporting period.	0	$0	$0
2. Audit reports on which management decisions were made during this period.	5	$0	$0
3. Audit reports on which final action was taken during this report period.			
(i) Disallowed costs that were recovered by management through collection, offset, property in lieu of cash, or otherwise.	0	$0	$0
(ii) Disallowed costs that were written off by management.	0	$0	$0
4. Reports for which no final action had been taken by the end of the reporting period.	5	$0	$0

Government Performance and Results Act: Review of the Fiscal Year 1999 Performance Report (OIG-01-A-03)

February 23, 2001

The Office of the Inspector General (OIG) of the U.S. Nuclear Regulatory Commission (NRC) conducted this audit at the request of the chairman of the Senate Committee on Governmental Affairs to determine whether NRC's fiscal year (FY) 1999 performance data were valid and reliable and whether the FY 2000 performance data would be more valid and reliable. The audit found that, while the NRC was improving and strengthening its performance reporting process, as interim policy guidance, the agency needed to institute management control procedures to produce valid and reliable data. The agency should then institutionalize the procedures in an NRC management directive (MD).

Open Recommendations	Actions Pending
1. Develop an NRC management directive (MD) to provide the management controls needed to ensure that the NRC produces credible Government Performance and Results Act (GPRA) documents.	The NRC issued interim guidance for performance management and reporting performance information in July 2001, consistent with GPRA requirements. Subsequently, the NRC issues agency guidance and instructions, annually, for completing GPRA documents, including reporting on unmet goals. The recommendations are currently in a "resolved" status.
3. Include guidance on reporting unmet goals in both the management directive and the interim policy guidance on implementing GPRA initiatives.	The recommendations will be addressed as part of the revision to the Management Directive (MD) and Handbook 4.7, "NRC Long Range Planning, Programming and Budget Formulation." We have modified our approach to the replacement of the MD and Handbook in order to improve policy communications, organization and achieve management consensus on the policies covered. We will replace it with three separate MDs: "Strategic Planning Process," "Budget Formulation," and, "Performance Management." Additionally, MD 4.4., "Management Controls," is currently being revised and re-titled, "Internal Controls." This MD establishes and assigns responsibility for controls and assurances over NRC programs and processes. The MD 4.7 replacement process is currently in a consultation phase.

Audit of the NRC's Regulatory Oversight of Special Nuclear Materials (OIG-03-A-15)

May 23, 2003

OIG conducted this audit to determine whether the NRC adequately ensures that its licensees control and account for special nuclear material (SNM). The audit found that NRC's current level of oversight of licensees' material control and accounting (MC&A) activities does not provide adequate assurance that all licensees properly control and account for SNM. The audit reported that the NRC performs only limited inspections of licensees' MC&A activities and thus cannot ensure the reliability of data in the Nuclear Materials Management & Safeguards System (NMMSS). The U.S. Department of Energy manages this computer database and shares it with the NRC as the national system for tracking certain private- and Government-owned nuclear materials.

Open Recommendations	Actions Pending
1. Conduct periodic inspections to verify that material licensees comply with MC&A requirements, including, but not limited to, visual inspections of licensees' SNM inventories and validation of report information. 3. Document the basis of the approach used to risk inform NRC's oversight of MC&A activities for all types of materials licensees.	In the February 7, 2006, memorandum, the Office of the Inspector General (OIG) stated that two of the three conditions identified by OIG that needed to be met to close this recommendation have been satisfied. The remaining condition is the need to complete documentation of the basis for riskinforming the MC&A program (and apply it to the program) with respect to conducting periodic inspections. In a subsequent memorandum dated August 24, 2006, OIG requested an estimated completion date for this recommendation. In SECY-05-0143, the staff recommended that the Commission approve the staffs proposed enhancements to the MC&A regulations, inspection program, and licensing process. Consistent with information provided in previous status reports, in response to the associated staff requirements memorandum (SRM) to SECY-05-0143 dated November 18, 2005, the staff completed the development of an MC&A rulemaking plan (SECY-08-0059) dated April 25, 2008. The SRM for SECY-08-0059 was issued on February 5, 2009. The Commission approved the staffs rulemaking Option 4, directing the staff to revise and consolidate current MC&A regulations into Part 74.

Audit of the NRC's Budget Formulation Process (OIG-05-A-09)

January 31, 2005

OIG conducted the audit to determine whether the budget formulation portion of the NRC's planning, budgeting, and performance management process is effectively used to develop and collect data to align resources with strategic goals and is efficiently and effectively coordinated with program and support offices. The audit found that the NRC effectively develops and collects data to align resources with strategic goals, prepares the budget in alignment with the Strategic Plan, and successfully conducts OMB-required program assessment rating tool evaluations. The audit also found that the agency needed additional internal coordination and communication efforts.

Open Recommendations	Actions Pending
1. Clarify the roles and responsibilities of the Chief Financial Officer and the Executive Director for Operations in the budget formulation process.	In August 2007, the Commission directed the Chief Financial Officer, in coordination with staff, to provide options for improving the agency's budget formulation process. The staff developed and implemented a new top-down budget process in formulating the agency's FY 2010 and FY 2011 budgets. Subsequently, the staff considered lessons learned from the NRC task force that reviewed the agency's budget formulation process.
2. Document the decisionmaking process and the roles and responsibilities of the program review committee.	
3. Document the budget formulation process to ensure a logical, comprehensive sequencing of events that provides for obtaining early Commission direction and approval.	Annually, the NRC Chairman issues guidance and budget instructions for developing and formulating the agency's budget which have incorporated improvements identified by the staff and task force. The guidance and instructions delineate the roles and responsibilities of the Chief Financial Officer and the Executive Director for Operations as well as others. The Program Review Committee has been eliminated with a more streamlined and efficient process. The guidance and instructions also document a logical, comprehensive sequence of events that provides for obtaining early Commission direction and approval. The recommendations are currently in a "resolved" status.

The recommendations will be addressed as part of the revision to the Management Directive (MD) and Handbook 4.7, "NRC Long Range Planning, Programming and Budget Formulation." We have modified our approach to the replacement of the MD and Handbook in order to improve policy communications, organization and achieve management consensus on the policies covered. We will replace it with three separate MDs: "Strategic Planning Process," "Budget Formulation," and, "Performance Management." The MD 4.7 replacement process is currently in a consultation phase. |

Audit of the NRC's Telecommunications Program (OIG-05-A-13)

June 7, 2005

OIG conducted this audit to evaluate controls over the use of NRC telecommunications services and the physical security of NRC telecommunications systems. OIG found that the agency needs to strengthen controls over the use of telecommunications services and the physical security of NRC telecommunications systems.

Open Recommendations	Actions Pending
3. Revise Management Directive 2.3 and Handbook, "Telecommunications," to include effective management controls over NRC headquarters staff use of agency telecommunications services.	The Office of Information Services (OIS) will submit the updated version of Management Directive 2.3, the written resolution of comments, and NRC Form 521 – "Request for Publication of an NRC Management Directive" to the Office of Administration Rules, Announcements and Directives Branch (ADM/RDB) for processing in early FY 2011.

UNITED STATES NUCLEAR REGULATORY COMMISSION

Audit of the NRC's Decommissioning Program (OIG-05-A-17)

September 21, 2005

OIG conducted this audit to determine whether the NRC's decommissioning program achieves desired performance results, as stated in the Strategic Plan and reported in the Performance and Accountability Report. The audit found that, while the NRC's decommissioning program has processes in place to monitor, evaluate, and report on performance, some performance results could not be verified. In addition, although staff implemented most of the recommendations from an FY 2003 self-evaluation of the program, the agency had not made progress on a few recommendations.

Open Recommendations	Actions Pending
1. Clarify and disseminate expectations for generating and maintaining supporting documentation for performance data to staff responsible for preparing and collecting performance data.	Annually, the NRC issues guidance for reporting performance data. The recommendation is currently in a "resolved" status. The recommendation will be addressed as part of the revision to the Management Directive (MD) and Handbook 4.7, "NRC Long Range Planning, Programming and Budget Formulation." We have modified our approach to the replacement of the MD and Handbook in order to improve policy communications, organization and achieve management consensus on the policies covered. We will replace it with three separate MDs: "Strategic Planning Process," "Budget Formulation," and, "Performance Management." "Performance Management" will address requirements for generating and maintaining supporting documentation for performance data. The MD 4.7 replacement process is currently in a consultation phase.

Audit of the NRC's Regulation of Nuclear Fuel Cycle Facilities (OIG-07-A-06)

January 10, 2007

This audit determined whether the NRC has an effective and efficient approach to fuel cycle facility oversight. The audit found that the NRC could enhance the current Fuel Cycle Facility Oversight Program by developing and implementing a framework modeled after a structured process, such as the Reactor Oversight Process (ROP).

Open Recommendations	Actions Pending
1. Fully develop and implement a framework for the Fuel Cycle Facility Oversight Program (FCFOP) that is consistent with a structured process, such as the Reactor Oversight Process (ROP).	Agency corrective actions include initiatives to improve fuel cycle oversight, including providing fuel cycle input to a revision of the NRC enforcement policy, and completing a safety culture pilot plan. The staff has drafted proposed changes to the NRC enforcement policy to align the policy with revisions to Title 10 of the Code of Federal Regulations (10 CFR) Part 70, "Domestic Licensing of Special Nuclear Material." The enforcement policy revision has been approved by the Commission and was issued September 30, 2010. The lengthiest corrective action is the two-phase Office of Nuclear Material Safety and Safeguards safety culture project plan, of which Phase I is complete. Phase II of the plan consists of implementing the Phase I results. The staff incorporated the Phase I results into the new FCOP which was rejected by the Commission in July 2010. The FCOP project has been modified in accordance with SRM-10-0031 in which the staff was directed to make modest adjustments to the existing oversight program to enhance its effectiveness and efficiency. Staff plans to incorporate the safety culture results into the revised FCOP as the project develops during in accordance with Commission direction and availability of resources.

Audit of Assessment of Security at NRC Buildings In Rockville, MD; Bethesda, MD; and Las Vegas, NV (OIG-07-A-18)

September 25, 2007

These security assessments determined the adequacy of physical security and emergency planning measures at the identified NRC buildings.

Open Recommendations	Actions Pending
11. Post signs near vehicle entrance directing pedestrians further west along Marinelli Avenue, and paint "Crosswalk" to direct pedestrians along a safe path to two controlled entry points.	Implementation of HSPD-12 included an overall assessment of physical access controls at the NRC headquarters complex. An NRC consultant completed an assessment of Recommendation 11 on February 29, 2008. Based on that assessment, the staff prepared a plan and cost analysis on installing a security fence to enclose the rear of the complex. The fence controls pedestrian traffic entering the One White Flint North and Two White Flint North buildings at the P1 levels. The installation of the fence was completed on August 20, 2010. This recommendation is considered to be completed. Closure is pending OIG review.

Audit of the NRC's Alternative Dispute Resolution Program (OIG-08-A-03)

December 14, 2007

This audit was conducted to determine whether the enforcement-related alternative dispute resolution (ADR) program, both early and postinvestigation ADR, was complete and ready for full implementation. The NRC deemed the ADR pilot program a success, and the staff, ADR participants, and other external stakeholders expressed satisfaction with the program. However, OIG found that the postinvestigation ADR process was not ready for full implementation because of weaknesses in the program's guidance and management controls.

Open Recommendations	Actions Pending
2. Incorporate the interim guidance into the Enforcement Policy and Manual.	The staff has incorporated ADR guidance, including guidance on the process for follow-up and closure of ADR confirmatory orders, in the revised Enforcement Policy which was published in the Federal Register (75 FR 60485) on September 30, 2010, (Reference ML093480037) with an immediately effective implementation date. ADR program guidance was also placed in the enforcement manual and was issued to the staff on December 22, 2008. This recommendation is considered to be completed. Closure is pending OIG review.

Audit of the NRC's Planned Cybersecurity Program (OIG-08-A-06)

March 18, 2008

This audit determined how upcoming changes to the NRC's cybersecurity oversight processes might impact the agency's physical security inspection program.

Open Recommendations	Actions Pending
1. Develop and implement plans for a cybersecurity oversight program that captures skill set and workload requirements for cybersecurity inspections, and targets resources to prepare for program implementation in calendar year 2010.	Proposed modifications were ranked as lower priorities than other activities in FY 2010. The staff has requested staff and contract resources for program development in calendar year 2011. With the requested resources, the staff plans to develop a temporary instruction inspection procedure and related enforcement guidance, conduct a pilot training course for the cyber security inspection team, conduct associated industry workshops, and conduct a pilot inspection. These actions will provide the framework for further development of the cyber security oversight program and the program's transition into the ROP.

Audit of the NRC's Continuity of Operations Plan (OIG-08-A-10)

May 21, 2008

This audit determined NRC's compliance with requirements for security surveys of the NRC's continuity of operations plan facilities.

Open Recommendations	Actions Pending
1. Revise current agency guidance governing security surveys of NRC continuity facilities to reflect Federal requirements (as originally stated in Federal Preparedness Circular 65 and superseded by Federal Continuity Directive 1) regarding annual physical security surveys of continuity facilities.	The revised MD 12.1, "NRC Facility Security Program," reflects the Federal Continuity Directive (FCD) 1, "Federal Executive Branch National Continuity Program and Requirements" requirement to provide for annual physical security inspections of continuity facilities. This document was sent to Directive Resources for publication on September 8, 2010.

Audit of the NRC's Accounting and Control Over Time and Labor Reporting (OIG-08-A-11)

June 17, 2008

OIG conducted an audit of the NRC's time and labor system on June 17, 2009. The objectives of the audit were to determine whether the NRC established and implemented internal controls over time and labor reporting to provide reasonable assurance that hours worked in pay status and hours absent are properly reported and that the time and labor system is easy and efficient to use.

Open Recommendations	Actions Pending
3. The CFO should conduct a detailed system analysis and eliminate redundant paper forms that are not needed.	The modernization project for the time and labor system is scheduled to be completed in the second Quarter of FY 2011. As part of this modernization, the Office of the Chief Financial Officer (OCFO) is working to incorporate an electronic workflow process, which would allow for electronic signatures. OCFO has met with the Office of Human Resources to discuss the possible elimination of various leave request forms and has also met with the National Treasury Employees Union. Preliminary findings indicated that the summary approval report, all leave request forms, unit transfer forms, and security request forms can be part of the electronic workflow process.
4. The CFO should ensure the use of electronic signature for time reporting and approval.	
	The modernization project for the time and labor system is scheduled to be completed in the second Quarter of FY 2011. As part of this modernization, OCFO is working to incorporate an electronic workflow process, which would allow for electronic signatures.

Audit of the NRC's Premium Class Travel (OIG-08-A-16)

September 12, 2008

OIG conducted an audit of the implementation of the NRC's premium class travel on September 15, 2008. The objectives of the audit were to determine whether travel costs associated with premium air travel (i.e., per diem) are properly authorized, justified, and documented and to determine whether premium air travel is properly authorized, justified, and documented. OIG specifically assessed compliance with requirements in OMB Memorandum M-08-07.

Open Recommendations	Actions Pending
1. Update Management Directive 14.1 to clearly identify premium travel authorizing officials; clarify "Delegation of Authority" and require this to be in written form; and clarify the 14-hour rule, specifically the rest period.	MD 14.1, "Official Temporary Duty Travel," has been revised to incorporate these changes. The staff is finalizing the various revisions and edits to the MD before it is submitted for formal review and concurrence. OCFO expects the MD 14.1 review and concurrence process to be completed, and MD 14.1 to be issued during FY 2011.

Audit of the NRC's Enforcement Program (OIG-08-A-17)

September 26, 2008

The objective of the audit was to review the NRC's enforcement program to determine whether the program is comprehensive and consistently implemented and whether enforcement decisions are based on complete and reliable data. OIG identified that the regional offices implement the enforcement program inconsistently because the agency has not issued clear and comprehensive guidance to facilitate the program. In addition, the audit identified that information used for decisionmaking and reporting purposes is not complete and reliable.

Open Recommendations	Actions Pending
2. Define data collection requirements for non-escalated actions.	The NRC staff is currently developing a Web-based licensing system that will track nonescalated enforcement actions issued to materials licensees. The database is expected to be available for enforcement data collection in mid 2011. The staff has evaluated the capabilities available with the reactor program system (RPS) and determined that it is a sufficient tool for tracking and trending nonescalated reactor enforcement actions.
3. Develop a quality assurance process to ensure that enforcement data is accurate and complete.	Actions to address Recommendation 3, which involve the development of procedures for data entry and auditing of WBL, will follow the actions to address Recommendation 2.

Implementation of the Federal Information Security Management Act for FY 2008 (OIG-08-A-18)

September 26, 2008

The objective of this review was to perform an independent evaluation of the Nuclear Regulatory Commission's (NRC) implementation of FISMA for fiscal year (FY) 2008.

Open Recommendations	Actions Pending
4. Develop a process for verifying that all Federal Desktop Core Configuration controls are implemented for all desktop and laptop computers, including both those that are centrally managed under the agency's seat management contract and those that are owned by the agency regardless of whether or not they are connected to the agency's network.	The staff will use Secure Content Automation Protocol (SCAP) and Federal Desktop Core Configuration (FDCC) compliance auditing tools to verify that the agency is compliant with M-08-22 for both OIS centrally managed and Region / Program Office managed computer assets.
	The staff will run these NIST-approved scanning tools against the Agency's image for standalone computers and against the agencies General Support Systems and Major Applications during system certification and accreditation and throughout continuous monitoring and quarterly security scanning, as required by FISMA.
	The SCAP and FDCC compliance tools will be part of the CSO Information Assurance System (IAS), which is scheduled to be deployed early Fiscal Year 2011.

Audit of the NRC's Committee to Review Generic Requirements (OIG-09-A-06)

February 2, 2009

The Office of the Inspector General (OIG) of the U.S. Nuclear Regulatory Commission (NRC) conducted this audit to determine if the Committee to Review Generic Requirements (CRGR) adds value for the Executive Director for Operations' decisionmaking purposes and whether the committee's function is still needed

Open Recommendations	Actions Pending
1. Develop, document, implement, and communicate an agencywide process for reviewing backfit issues to ensure that generic backfits are appropriately justified based on NRC regulations and policy.	In addressing Recommendation 1 and in its role of providing CRGR support, the staff coordinated the implementation of an Action Plan with the relevant offices and regions. The planned activities are currently envisioned to include at least the following five areas: (1) revise the CRGR Charter, (2) revise Management Directive (MD) 8.4, "Management of Facility-specific Backfitting and Information Collection", (3) develop office and regional procedures that are consistent with the revised MD 8.4, (4) develop an agencywide Web-based backfit training program, and (5) document, communicate, and implement an overarching agencywide backfit program. The CRGR and Office staff worked together to establish a centralized agency resource for backfit training.
	At the present, CRGR and Office staff are in the process of reviewing and updating a previous draft of an agencywide Web-based backfit training. The next step will be to develop a training module on the overall process and then to develop program-specific modules that can be used by the program offices and regions, as appropriate.
	The NRC is presently revising the CRGR charter and various NRC offices are coordinating to revise the MD 8.4 to reflect changes in NRC's organizational responsibilities and backfit program. These revisions will address important elements for ensuring effective overarching management of generic and plant-specific backfits.
	These planned activities will document the role of the CRGR and the staff process for ensuring compliance with backfit requirements and procedures that have evolved since the inception of the CRGR. The CRGR plans to communicate the changes to the staff and verify that the relevant offices and regions have incorporated processes to ensure backfit rules and requirements are followed.

Audit of NRC's Occupant Emergency Program (OIG-09-A-07)

February 11, 2009

The audit objective was to evaluate the extent to which the agency's Occupant Emergency Program complies with Federal regulations and standards.

Open Recommendations	Actions Pending
2. Require annual, unannounced, full-scale evacuation drills, including mustering and accountability assessments, at all headquarters and regional complexes.	Full-scale emergency evacuation drills, including assembly and accountability (A&A) were conducted in calendar year 2009 at all headquarters and regional complexes, with the exception of Region II. In August 2010, the building management for the complex in which the Region II office is located conducted training for the NRC floor and stairwell monitors in preparation for an upcoming fire drill. Region II moved into the complex on April 12, 2010, and a fire drill has been postponed by the building manager due to several new tenants moving into the complex. Region II was recently advised by the building manager that an unannounced fire drill will be conducted in early FY 2011. Region II will conduct an A&A of staff during the fire drill evacuation.

Audit of the NRC's Agreement State Program (OIG-09-A-08)

September 28, 2010

The audit objective was to assess NRC's oversight of the adequacy and effectiveness of Agreement State programs. OIG focused its review on the IMPEP process as well as other elements of the Agreement State program. OIG identified program adequacy and effectiveness issues that require management's attention.

Open Recommendations	Actions Pending
1. Develop mechanism for conducting self-assessments and capturing lessons learned for IMPEP on a regular basis.	A review team of NRC and Agreement State staff completed the first self-assessment of the IMPEP program in June 2010. The self-assessment report included a draft procedure for conducting future self-assessments. FSME will implement the 15 recommendations and enhancements made in the self-assessment.
2. Develop formal procedural guidance for identifying what information is needed about Agreement State programs and materials licensees in the event that an Agreement State is no longer capable of adequately performing its function of protecting public health and safety for an indeterminate period of time.	The NRC staff is revising FSME State Agreement procedure SA-114, "Suspension of a Section 274b. Agreement," to address this recommendation.
3. Develop a set of procedures that standardizes communication s from NRC to the Agreement States.	The NRC staff is revising FSME procedure AD-200, "Format for FSME Letters" to address this recommendation.
4. Develop a standardized data collection process that can be used as the basis of an information sharing tool on a national level.	NRC staff sent a questionnaire to the Agreement States in April 2010 to obtain their input on their willingness to share certain enforcement and allegation information, estimates on the annual burden to share the information, and the legality of sharing the information from the State perspective. Approximately one third of the Agreement States responded. FSME will be contacting more States in early FY 2011 prior to making a recommendation on a data collection process for the information.

Audit of NRC's Warehouse Operations (OIG-09-A-09)

March 31, 2010

The purpose of this audit was to determine whether NRC has established and implemented an effective system of internal controls for maintaining accountability and control of agency property stored in the warehouses.

Open Recommendations	Actions Pending
2. Conduct the required security survey of the NRC annex.	The Federal Protective Service Area Commander notified the Office of Administration (ADM), Division of Facilities and Security, that a Building Security Assessment of the NRC Annex was completed on August 6, 2010. The NRC was advised that a final security survey report will be issued approximately 45 days after the assessment completion date. Upon receipt, review and approval of the report by ADM, a copy will be sent by ADM to the OIG. ADM will then consider this recommendation to be completed.

Information System Security Evaluation of the Technical Training Center - Chattanooga, TN (OIG-09-A-11)

July 22, 2009

The Office of the Inspector General (OIG) of the U.S. Nuclear Regulatory Commission (NRC) conducted this audit pursuant to the Federal Information Security Management Act (FISMA) of 2002. The FISMA requires an annual independent evaluation of an agency's information security program and practices to determine their effectiveness. This audit evaluates the information security policies, procedures, and practices at the agency's Technical Training Center (TTC), which was last assessed in 2003 and 2006. The audit found that TTC has made improvements in its implementation of NRC's information system security program since previous audits. While many of the TTC's automated and manual security controls were found to be generally effective, some security controls were found to need improvement

Open Recommendations	Actions Pending
Recommendations were provided to improve some security controls.	The staff has completed 1 of the 3 open recommendations and its closure is pending OIG review. The remaining 2 open recommendations are planned to be completed by in early FY 2011.

Office of the Inspector General Information System Evaluation of Region II - Atlanta, Georgia (OIG-09-A-13)

September 28, 2009

OIG requested that the Region II office be included in the independent evaluation of the agency's implementation of FISMA for fiscal year 2009, with the objectives of evaluating the Region's information security program and practices to determine their effectiveness; including related information security policies, procedures, standards, and guidelines. The audit found that while many of the Region II automated and manual security controls are generally effective, some security controls needed improvement.

Open Recommendations	Actions Pending
Recommendations were provided to improve some security controls.	Region II has completed all the recommendations, and closure of the recommendations is pending OIG review.

Office of the Inspector General Information System Evaluation of Region IV – Arlington, Texas (OIG-09-A-14)

September 28, 2009

OIG requested that the Region IV office be included in the independent evaluation of the agency's implementation of FISMA for fiscal year 2009 in order to evaluate the Region's information security program and practices. The results of audit found that while many of Region IV's automated and manual security controls are generally effective, some security controls needed improvement.

Open Recommendations	Actions Pending
Recommendations were provided to improve some security controls.	Region IV has completed 3 of the 6 OIG recommendations, and closure is pending OIG review.

Office of the Inspector General Information System Evaluation of Region III – Lisle, IL (OIG-09-A-15)

September 28, 2009

OIG requested that the Region III office be included in the independent evaluation of the agency's implementation of FISMA for fiscal year 2009, with the objectives of evaluating the Region's information security program and practices to determine their effectiveness; including related information security policies, procedures, standards, and guidelines. The audit found that while many of the Region III automated and manual security controls are generally effective, some security controls needed improvement.

Open Recommendations	Actions Pending
Recommendations were provided to improve some security controls.	Region III has completed all the OIG recommendations, and closure of 4 of the recommendations is pending OIG review.

Audit of the NRC's Grant Management Program (OIG-09-A-16)

September 29, 2009

The audit objective was to determine whether NRC has established and implemented an effective system of internal controls for grants management.

Open Recommendations	Actions Pending
1. Resolve outstanding Lean Six Sigma issues, including definition of the competitive grant process, roles and responsibilities, development of a shared electronic grant database, and scope of Office of Small Business and Civil Rights reviews.	The Office of Administration resolved this recommendation in part by incorporating Lean Six Sigma recommendations in interim guidance for the grants program to include the definition of the competitive grants process and a section on roles and responsibilities. The interim guidance was issued as a draft of Management Directive (MD) 11.6, "Financial Assistance Program," on May 28, 2010, to the Office of Human Resources, Office of Nuclear Regulatory Research, Office of the General Counsel and the Office of Small Business and Civil Rights. With respect to the development of a shared electronic database, ADM currently stores financial assistance applications in an electronic, shared, financial assistance folder that is accessible to program office personnel involved in the financial assistance process. This part of Recommendation 1 will be resolved through the development of a SharePoint site for grants management by in early FY 2011, which will include an improved document/reference library. In addition, ADM will continue to coordinate with the Office of the Chief Financial Officer as it develops potential functionality and capabilities within the Nuclear Regulatory Commission's Financial Accounting and Integrated Management System (FAIMIS) to confirm whether the planned grants module will support a grants database.
2. Update Management Directive 11.6 to comprehensively address NRC's competitive and non-competitive grant program, including (a) roles and responsibilities of individuals and offices involved in the grant process, (b) process for awarding grants, and (c) required monitoring by project officers.	ADM is currently updating MD 11.6 through the formal MD process to provide consistent policies and procedures for awarding, administering and monitoring competitive and noncompetitive grants, and to clarify the roles and responsibilities of ADM Division of Contracts (DC) and program office personnel involved in the process. ADM issued MD 11.6, "Financial Assistance Program" for formal office comment on July 23, 2010.
5. Ensure that staff working on grants complete the required training within the specified timeframe identified in response to recommendation 4.	Grant staff must complete the required training identified in NRC's Grant Management Certification Training Program by December 31, 2011. This training is being monitored by the NRC Acquisition Career Manager in coordination with the Office of Human Resources.

6. Develop a method for sharing up-to-date official file/grant documentation with all involved parties to include a formal electronic tracking and reporting system.	ADM's Automated Acquisition Management System provides access to grant award documents by staff in the Division of Contracts and program offices involved in the grants process. This recommendation will be resolved through the development of a SharePoint site for grants management which will include an improved document/reference library. In addition, ADM will continue to coordinate with the Office of the Chief Financial Officer as it develops potential functionality and capabilities within FAIMIS to confirm whether the planned grants module will support a grants database.
8. Develop a quality assurance program for ensuring official grant files are complete	Contract number NRC-10-08-373, an 8(a) contract, which provided for independent file reviews of commercial contracts, Department of Energy Laboratory Agreements, and other interagency agreements, expired on July 31, 2010. ADM included the requirement to develop a quality assurance process in the Statement of Work for the new contract under Request for Proposal Number ADM-10-397, which was awarded in August 2, 2010.
	In support of the new quality assurance process, the contractor will develop a checklist to ensure the accuracy and adequacy of grant files, determine if all appropriate procedures were followed, and provide a list of missing documents.

Audit of NRC's Oversight of Construction at New Nuclear Facilities (OIG-09-A-17)

September 29, 2009

OIG conducted this evaluation to determine if and how NRC is identifying and incorporating lessons learned in its new Construction Inspection Program.

Open Recommendations	Actions Pending
1. Enhance CIP guidance, which includes NRO-REG-112, to include key elements identified as important to the success of an organization's lessons learned program. Specifically:	The staff is on track to revise NRO-REG-112 by the end of calendar year 2010 to incorporate all OIG recommendations as described in OIG-09-A-17.
a. Define "lessons learned" as it applies to new reactor construction.	
b. Establish and document collection criteria for the types of information that CIP staff should bring forward for screening as potential lessons learned.	
c. Further develop and document how a construction-related lesson learned will be implemented through the CIP.	
d. Establish and document the level of expertise required for staff participation in the daily screening meetings.	

Audit of NRC's Material Control and Accounting Security Measures for Special Nuclear Materials at Fuel Cycle Facilities (OIG-09-A-19)

September 30, 2009

Open Recommendations	Actions Pending
1. Review and revise MC&A procedures as required by NMSS Policy and Procedures Letter 1-76.	The MC&A inspection procedures (IPs) are currently being reviewed and revised, as necessary. As of September 23, 2010, the first six IPs have been revised and issued by NRR. Furthermore, nine additional IPs are in the final concurrence stage and Inspection Manual Chapter 2683 (Material Control and Accounting [MC&A] Inspection Program for Fuel Cycle Facilities) was revised and submitted to Region II for comments as part of this inspection procedure update process. In order to ensure completion of the active Category I and Category Ill IPs, NMSS has deferred action on the Cat II procedures as an effectiveness and efficiency measure. Delaying the revision of the Category II inspection procedures will have no impact on the current NMSS MC&A inspection program activities or effectiveness, in that there are no Category II facilities to which these procedures currently apply. Therefore, NMSS will wait for the completion of the current Part 74 rulemaking activities before determining whether to expend resources updating the Category II inspection procedures unless there are significant delays in the rulemaking and a license application for a Category II is received.
2. Establish an alternative to DOE-sponsored MC&A inspector training to be used as needed.	On July 26, 2010, NMSS established a contract with DOE's National Training Center (NTC) to develop training materials that can be used as self-study guides to replace the course content in each DOE-sponsored MC&A and related class that has not been routinely offered. To date, NMSS has received two of the five sets of course materials. The Technical Project Manager and the NMSS Lead MC&A Inspector are currently reviewing these course materials.

In addition, NTC management has advised U.S. Nuclear Regulatory Commission (NRC) staff that the minimum class requirements have been significantly reduced and the NTC sponsored classes, which have been problematic for NMSS staff to attend in the past, will therefore be available on a more frequent basis. While this information is useful and encouraging, NMSS is proceeding with establishing the NRC's own training materials via the aforementioned contract. |

Information System Security Evaluation of Region I – King of Prussia, PA (OIG-09-A-20)

September 30, 2009

OIG conducted this evaluation to evaluate the adequacy of NRC's information security program and practices for NRC automated information systems as implemented at Region I; evaluate the effectiveness of agency information security control techniques as implemented at Region I; and evaluate corrective actions planned and taken as a result of previous OIG evaluations.

Open Recommendations	Actions Pending
Recommendations were provided to improve some security controls.	Region I has completed all the recommendations, and closure of some recommendations is pending OIG review.

Summary of Financial Statement Audit and Management Assurances

Jan Kweiser and Justine Burza, winners of the Chicago 2010 Federal Executives of the Year Awards – May 2010.

Photo Courtesy of NRC Photo Library

Region IV staff in the Incidence Response Center during an emergency response exercise – April 2010.

Summary of Financial Statement Audit and Management Assurances

SUMMARY OF FINANCIAL STATEMENT AUDIT

Audit Opinion—Unqualified

Restatement—No

Material Weaknesses—No

SUMMARY OF MANAGEMENT ASSURANCES

Effectiveness of Internal Control over Financial Reporting (FMFIA § 2)

Statement of Assurance—Unqualified

Material Weaknesses—No

Effectiveness of Internal Control over Operations (FMFIA § 2)

Statement of Assurance—Unqualified

Material Weaknesses—No

Conformance with Financial Management System Requirements (FMFIA § 4)

Statement of Assurance—Systems Conform to Financial Management System Requirements

Nonconformance—No

Compliance with Federal Financial Management Improvement Act (FFMIA)

	Agency	Auditor
Overall Substantial Compliance	Yes	Yes
1. Systems Requirements	Yes	Yes
2. Accounting Standards	Yes	Yes
3. United States Standard General Ledger at Transaction Level	Yes	Yes

Photo Courtesy of NRC Photo Library

NRC staffers John Ellegood and Jack Geissner receive Team Player Awards from Region III Administrator, Mark Satorius, and Executive Director for Operations, R. William Borchardt – June 2010.

Acronyms and Abbreviations

Photo Courtesy of NRC Photo Library

Diablo Canyon Nuclear Power Plant,
San Luis Obispo County, CA.

Acronym	
10 CFR	Title 10 of the *Code of Federal Regulations*
ADAMS	Agencywide Documents Access and Management System
ADM	Office of Administration
ADR	alternative dispute resolution
ALC	agency location code
C&A	certification and accreditation
CCDP	Conditional core damage probability
CFO	Chief Financial Officer
CFR	*Code of Federal Regulations*
CoC	Certificates of Compliance
COL	Combined Operating License
CRGR	Committee to Review Generic Requirements
CSO	Computer Security Office
CSRS	Civil Service Retirement System
CUI	controlled unclassified information
DHS	Department of Homeland Security
DOE	U.S. Department of Energy
DOI-NBC	Department of the Interior National Business Center
DOL	U.S. Department of Labor
ECIC	Executive Committee on Internal Control
EDO	Executive Director for Operations
e-Gov	Federal Government's Electronic Government
e-OPF	electronic official personnel folders
EPR	Evolutionary Power Reactor
ESBWR	Economic Simplified Boiling-Water Reactor
FAIMIS	Financial Accounting and Integrated Management System
FBI	Federal Bureau of Investigation
FCD	Federal Continuity Directive
FCFOP	Fuel Cycle Facility Oversight Program
FDCC	Federal Desktop Core Configuration
FECA	Federal Employees Compensation Act
FEMA	Federal Emergency Management Agency

Acronym	
FERS	Federal Employees Retirement System
FFLI	Fuel Facilities Licensing and Inspection
FFMIA	Federal Financial Management Improvement Act
FICA	Federal Insurance Contribution Act
FISMA	Federal Information Security Management Act
FMFIA	Federal Managers' Financial Integrity Act
FOIA	Freedom of Information Act
FR	*Federal Register*
FY	fiscal year
GAAP	generally accepted accounting principles
GALL	Generic Aging Lessons Learned
GEIS	generic environmental impact statement
GEM	graphical evaluation module
GPRA	Government Performance and Results Act
GSA	General Services Administration
HSPD	Homeland Security Presidential Directive
HTGR	high-temperature gas-cooled reactor
I-131	Iodine-131
IAS	Information Assurance System
IAEA	International Atomic Energy Agency
IG	Inspector General
IMPEP	Integrated Materials Performance Evaluation Program
Improvement Act	Federal Financial Management Improvement Act
Integrity Act	Federal Managers Financial Integrity Act
IPCE	Integrated Pilot Comprehensive Exercise
IPs	inspection procedures
IPSS	Integrated Personnel Security System
iPWR	Integral pressurized-water reactor
ISA	Integrated safety analysis
ISG	interim staff guidance
IT	information technology

Acronym	
ITAAC	inspections, tests, analyses, and acceptance criteria
LES	light-water facilities
LWR	graphical evaluation module
MC&A	material control and accounting
MD	management directive
MDEP	Multinational Design Evaluation Program
MOX	Mixed Oxide
NEA	Nuclear Energy Agency
NEI	Nuclear Energy Institute
NERC	North American Electric Reliability Corporation
NMMSS	Nuclear Materials Management and Safeguards System
NOV	notices of violation
NRC	U.S. Nuclear Regulatory Commission
NSTS	National Source Tracking System
NTC	National Training Center
NUREG	Nuclear Regulatory Commission document identifier
NWF	Nuclear Waste Fund
OBRA-90	The Omnibus Budget Reconciliation Act of 1990
OCWE	Open Collaborative Working Environment
OI	Office of Investigation
OIG	Office of the Inspector General
OIS	Office of Information Services
OMB	Office of Management and Budget
OUO	official use only
PAR	Performance and Accountability Report
PII	personally identifiable information

Acronym	
POA&M	plan of action and milestones
PRA	probabilistic risk assessment
PTS	pressurized thermal shock
REM	Roentgen Equivalent Man
ROP	Reactor Oversight Process
RPS	reactor program system
SAPHIRE	Systems Analysis Program for Hands-On Integrated Reliability Evaluations
SCAP	Secure Content Automation Protocol
SECY	Office of the Secretary of the Commission
SFFAS	Statement of Federal Financial Accounting Standards
SGI	safeguards information
SMR	small modular reactor
SNM	special nuclear material
SOARCA	State-of-the-Art Reactor Consequences Analysis
SRM	staff requirements memorandum
SSEP	safety, security, and emergency preparedness
SUNSI	sensitive unclassified, nonsafeguards information
T&L	time and labor
TAD	transportation, aging, and disposal
TSP	Thrift Savings Plan
TTC	Technical Training Center
TVA	Tennessee Valley Authority
UO2	Uranium Dioxide
USAID	U.S. Agency for International Development
USEC	United States Enrichment Corporation
USSP	United States Support Program
V&V	verification and validation

NRC FORM 335 (9-2004) NRCMD 3.7	U.S. NUCLEAR REGULATORY COMMISSION	1. REPORT NUMBER (Assigned by NRC, Add Vol., Supp., Rev., and Addendum Numbers, if any.)
	BIBLIOGRAPHIC DATA SHEET *(See instructions on the reverse)*	NUREG-1542, Vol. 16

2. TITLE AND SUBTITLE	3. DATE REPORT PUBLISHED	
U.S. Nuclear Regulatory Commission FY 2010 Performance and Accountability Report	MONTH	YEAR
	November	2010
	4. F N OR GRANT NUMBER	
	N/A	

5. AUTHOR(S)	6. TYPE OF REPORT
David Holley, et. al	Annual
	7. PERIOD COVERED *(Inclusive Dates)*
	Fiscal Year 2010

8. PERFORM NG ORGANIZATION - NAME AND ADDRESS *(If NRC, provide Division, Office or Region, U.S. Nuclear Regulatory Commission, and mailing address; if contractor, provide name and mailing address.)*

Division of Planning and Budget
Office of the Chief Financial Officer
U.S. Nuclear Regulatory Commission
Washington, DC 20555-0001

9. SPONSOR NG ORGANIZATION - NAME AND ADDRESS *(If NRC, type Same as above ; if contractor, provide NRC Division, Office or Region, U.S. Nuclear Regulatory Commission, and mailing address.)*

Same as 8, above

10. SUPPLEMENTARY NOTES

11. ABSTRACT *(200 words or less)*

The Fiscal Year 2010 Performance and Accountability Report (PAR) presents the agency's program performance and financial management information. The PAR gives the President, Congress, and the American public the opportunity to assess the agency's performance in achieving its mission, and the stewardship of its resources.

12. KEY WORDS/DESCRIPTORS *(List words or phrases that will assist researchers in locating the report.)*	13. AVAILABILITY STATEMENT
Performance and Accountability Report (PAR) Fiscal Year (FY) 2010	unlimited
	14. SECURITY CLASSIFICATION
	(This Page) unclassified
	(This Report) unclassified
	15. NUMBER OF PAGES
	16. PRICE

NRC FORM 335 (9-2004)

PRINTED ON RECYCLED PAPER